Apollo Advanced Lunar Exploration Planning

Edited and Compiled from the NASA files
by Robert Godwin

An Apogee Books Publication

All rights reserved under article two of the Berne Copyright Convention (1971).
We acknowledge the financial support of the Government of Canada through the Book
Publishing Industry Development Program for our publishing activities.

Published by Apogee Books, Box 62034, Burlington,
Ontario, Canada, L7R 4K2, http://www.apogeebooks.com
Tel: 905 637 5737

Printed and bound in Canada
Apollo Advanced Lunar Exploration Planning
Layout ©2007 Robert Godwin
All images courtesy NASA
Compiled and Edited by Robert Godwin
ISBN 978-1-894959-80-3

APOLLO ADVANCED LUNAR EXPLORATION PLANNING

1970 Compiled by Lunar Missions Office Advanced Spacecraft Technology Division

CONTENTS

Section	Page
I INTRODUCTION	11
II SCHEDULES	12
III MANNED LANDER DESCRIPTION	14
Lunar Module	14
Extended Lunar Module	15
Augmented Lunar Module	15
Definitions	15
IV UNMANNED LANDER DESCRIPTION	19
Lunar Payload Module	19
Lunar Module Truck	19
Direct Delivery of Unmanned Landers	20
V LUNAR ORBIT MISSION CAPABILITY	21
Lunar Mapping and Survey System	21
Manned and Unmanned Lunar Landings	22
VI AUXILIARY SHELTER AND TRUCK SHELTER VEHICLES	24
Lunar Module Shelter	24
Early Lunar Shelter	25
Staytime Extension Module	26
Lunar Module Truck Shelter	26
Lunar Shelter Comparison	27
VII PAYLOAD CAPABILITIES	31
Surface Payload	31
Orbital Payloads	32
VIII OPERATIONAL CONSIDERATIONS	36
Flight Times and Mission Duration	36
Multiple-Burn Lunar Orbit Insertion	36
Landing Flight Modes	37
Landing Accuracies	37
Landing Site Latitude and Longitude Effects	37
Surface-Staytime	38
Abort Considerations	38
Lunar Surface Lighting	39
Mobility Aids	39
Communications	40

IX
STAYTIME EXPENDABLES — 44
- Expendables Versus Days Staytime — 45
- Life Support Requirements — 46
- Electrical Power Generation — 46

X
LUNAR MOBILITY — 48
- Lunar Roving Vehicles (Manned) — 48
- Local Scientific Survey Module — 48
- Mobile Exploration, Mobile Laboratory Vehicles — 49
- Lunar Flying Machines — 50

XI
LUNAR EXTRAVEHICULAR ACTIVITY OPERATIONS — 56
- Space Suits — 57
- Life Support Systems — 58
- Working in Lunar Environment — 59

XII
ASTRONAUT LUNAR TIME LINES — 65
- Time Required for Mission Operations — 65
- Time Required for Normal Life Sustenance — 65
- Time Available for Experiments and Exploration — 65

XIII
EXPERIMENT MODULES — 67
- Apollo Lunar Surface Experiments Package — 67
- ALSEP Experiments — 68
- Lunar Surface Drill — 68
- Deep Drill for Lunar Surface — 69
- Lunar Survey System Staff Device — 70
- Lunar Mapping and Survey System — 71

XIV
EXPERIMENT EQUIPMENT REQUIREMENTS — 74
- Equipment Availability Requirements — 74
- Equipment Qualification Requirements — 75
- Training Requirements — 75

XV
LAUNCH VEHICLE CHARACTERISTICS — 76
- Payload Capability — 76
- Kennedy Spacecraft Center Launch Operations — 78

XVI
APOLLO COMMAND AND SERVICE MODULE — 83
- Command Module — 83
- Service Module — 83
- Spacecraft Lunar Module Adapter — 83

XVII
UNMANNED LUNAR SATELLITES AND PROBES — 85
- Satellite Delivery by Apollo — 85
- Landing Probes — 86
- Surface Crawlers — 86

XVIII
HARDWARE AND MISSION COSTS. — 90

TABLES

VI-I LUNAR SHELTER COMPARISONS	27
VII-I LUNAR SURFACE PAYLOADS	33-34
VII-II LUNAR-ORBITAL MISSION PAYLOAD	34
VIII-I LUNAR MODULE LANDING DISPERSION SUMMARY	40
X-I FLYING VEHICLES INVESTIGATED	51
X-II ONE-MAN FLYING MACHINE	51
XI-I SPACE SUIT WEIGHT COMPILATION	60
XI-II APPROXIMATE PERCENT OF NUDE JOINT MOBILITY RANGE RETAINED AT 3.7 PSIG	60
XI-III HARD SUIT ASSEMBLY WEIGHT COMPILATION	61
XI-IV APPROXIMATE PERCENT OF NUDE JOINT MOBILITY RANGE RETAINED AT 3.7 PSIG	61
XVII-I CAPABILITIES OF SATELLITES	87
XVIII-I NONRECURRING AND ITEM COST SUMMARY	91
XVIII-II GROUP 1 MISSIONS ($1095.7 MILLION)	92
XVIII-III GROUP 2 MISSIONS ($1503.1 MILLION)	93
XVIII-IV GROUP 3 MISSIONS ($1627.5 MILLION)	94
XVIII-V GROUP 4 MISSIONS ($1675.2 MISSION)	95

FIGURES

II-1 Availability and use periods of lunar spacecraft	14
III-1 Lunar module configuration	16
III-2 Lunar module overall dimensions	17
III-3 Lunar module ascent stage	18
III-4 Lunar module descent stage	18
IV-1 Lunar module truck concepts	
(a) LM ascent stage	20
(b) Integrated LM truck	20
(c) Modular LM truck	20
V-1 Service module, block II	22
V-2 The anchored interplanetary monitoring platform	23
V-3 Orbital experiment module launch configuration	23
V-4 Orbital experiment module lunar orbit configuration	24
VI-1 Lunar module shelter	
(a) Front view	27
(b) Inboard profile	27
VI-2 Lunar module shelter work area	
(a) Top view	28
(b) Side view	28
(c) End view	28
VI-3 Sleeping arrangement concept	28
VI-4 Sleeping arrangement concept, forward cabin	29
VI-5 Early lunar shelter	
(a) Side view	29
(b) Top view,	29
VI-6 Early lunar shelter, detail view	
(a) End view	30
(b) Side view	30
VI-7 Staytime extension module	30

VI-8 Lunar module truck shelter	31
VII-1 Usable volume of the lunar module	
(a) Launch configuration	35
(b) Landed configuration	35
(c) Plan view of descent stage	35
(d) Plan view of ascent stage	35
VIII-1 Three-impulse geometry for lunar orbit insertion	41
VIII-2 Lunar landing accessibility using the 14-day command and service module	41
VIII-3 Lunar landing accessibility using the 14-day command and service module, as viewed from earth	42
VIII-4 Lunar landing accessibility using the 23-day command and service module, as viewed from earth	43
VIII-5 Lunar orbit geometry for staytime, site latitude, and abort considerations	44
IX-1 Expendables versus days staytime	47
IX-2 Weight trade-off airlock/no airlock versus number of EVA sorties	47
IX-3 Modular mission extension provisions	48
X-1 Local scientific survey module concept, four wheel	52
X-2 Local scientific survey module concept, six wheel	52
X-3 Lunar module shelter, or stripped lunar scientific module stowed configuration	53
X-4 Mobile exploration, mobile laboratory type vehicles	53
X-5 Lunar flying unit	
(a) Top view	54
(b) Front view	54
X-6 Manned flying vehicle	54
X-7 Flight time versus distance	55
X-8 Flyer weight versus distance and propellant (manned)	55
X-9 Flyer velocity versus distance	56
XI-1 Liquid coolant garment	62
XI-2 Pressure garment assembly	62
X1-3 Thermal meteoroid garment	63
XI-4 Hard suit, front view	63
XI-5 Hard suit, side view showing backpack	64
XI-6 Times to conduct elementary tasks	64
XIII-1 Apollo lunar surface experiments package telemetry system	72
XIII-2 Apollo lunar surface experiments package delivery system	72
XIII-3 Apollo lunar surface experiments package array A (typical)	73
XIII-4 Apollo lunar surface experiments package array B (typical)	73
XIII-5 Lunar survey system staff	74
XV-1 Saturn V launch vehicle and Apollo spacecraft	77
XV-2 Space vehicle flow	80
XV-3 Saturn V countdown and hold decision points	81
XV-4 Saturn V turnaround time from scrub to next T-0	81
XV-5 Space vehicle reschedule plan	82
XV-6 Mobile service structure flow plan, two-vehicle launch	82
XVI-1 Apollo spacecraft	84
XVI-2 Command module	84
XVI-3 Service module	85
XVII-1 Surveyor probe	87
XVII-2 Typical hard lander erection and. leveling sequence proposal I	88
XVII-3 Typical hard lander erection and leveling sequence, proposal II	88
XVII-4 Soft lander probe	89
XVII-5 Unmanned lunar surface crawler	89

ACRONYMS

AAP	Apollo Applications Program
ACE	acceptance checkout equipment
AIMP	anchored interplanetary monitoring platform
ALM	augmented lunar module
ALSEP	Apollo lunar surface experiments package
APS	auxiliary propulsion system
ARC	Ames Research Center
AS	ascent stage
ASE	active seismic experiment
CCGE	cold cathode gage experiment
CD	countdown
CDDT	countdown demonstration test
CM	command module
CPLEE	charged-particle lunar environment equipment
CSM	command and service module
CT	crawler transporter
CWG	constant wear garment
DS	descent stage
ECS	environmental control system
ELM	extended lunar module
ELS	early lunar shelter
Eng	engineering
EPS	electrical power subsystem
ERA	electronic replaceable assembly
EVA	extravehicular activity
FRT	flight readiness test
GFE	Government-furnished equipment
GRCSW	Graduate Research Center of the Southwest
GSE	ground support equipment
GSFC	Goddard Space Flight Center
He	helium
HFE	heat flow experiment
IMU	inertial measurement unit
IR	infrared
IRU	inertial reference unit
KSC	Kennedy Space Center
LCG	liquid coolant garment
LH2	liquid hydrogen
LM	lunar module
LMS	lunar module shelter
LM&SS	lunar mapping and survey system
LMTS	lunar module truck shelter
LN2	liquid nitrogen
LO2	liquid oxygen
LPM	lunar payload module
LSC	lunar surface crawler
LSM	lunar surface magnetometer
LSSM	local scientific survey module
LV	launch vehicle
MCC-H	Mission Control Center, Houston
MFS	manned flying system
MOBEX	mobile excursion

MOLAB	mobile laboratory
MSFN	Manned Space Flight Network
MSOB	Manned Spacecraft Operations Building
MSS	mobile service structure
NASA	National Aeronautics and Space Administration
OEM	orbital experiment module
PDA	predelivery acceptance test
PECS	portable environmental control system
PGA	pressure garment assembly
PIA	. preinstallation acceptance test
PLSS	portable life support system
PSE	passive seismic experiment
RCS	reaction control system
RA	rocket propellant
RTG	radioisotope thermoelectric generator
RTTG	radioisotope thermionic generator
S&A	safe and arm
SC	spacecraft
Sci	Scientific
SFT	simulated flight test
SHe	supercritical helium
SIDE	suprathermal ion detector experiment
S-IC	Saturn IC
S-II	Saturn II
S-IVB	Saturn IVB
SLA	spacecraft lunar module adapter
SM	service module
SNAP	system for nuclear auxiliary power
SPS	service propulsion system
STEM	staytime extension module
S-V	Saturn V
SWS	solar-wind spectrometer
TMG	thermal meteoroid garment
TV	television
UHF	ultrahigh frequency
VAB	Vertical Assembly Building
VHF	very high frequency

DEFINITIONS

AAP — Apollo Application Program.- The AAP includes the post Apollo efforts using uprated Apollo hardware.

AIMP — Anchored interplanetary monitoring platform.- The AIMP, a Goddard Space Flight Center design, is a 210-pound remote-sensor satellite.

ALM — Augmented lunar module.- The ALM is a lunar landing spacecraft with an uprated descent stage to land 2000 pounds more payload.

ALSEP — Apollo lunar surface experiments package.- The ALSEP is a 225pound instrument package to be deployed for operation on the lunar surface for 1 year.

CM — Command module.- The CM is the Apollo spacecraft crew compartment.

CSM — Command and service module.- The CSM is the Apollo spacecraft including the cabin, the propulsion stage, and the expendables.

CT — Crawler transporter.- The CT is a power drive unit for transporting the Saturn V Apollo spacecraft assembly from the checkout facility to the launch pad.

CWG	Constant wear garment.- The CWG is the suit used inside the Apollo spacecraft for emergency pressure provisions.
ELM	Extended lunar module.- The ELM is the Apollo lunar lander with an increased capability derived through changes in Apollo operational techniques.
ELS	Early lunar shelter.- The ELS is a hard-shelled shelter structure for habitation and work on the lunar surface.
EPS	Electrical power subsystem.- The EPS includes batteries, fuel cell, and/or solar arrays, controls, and wiring.
EVA	Extravehicular activity.- The EVA is the suited-astronaut activity outside of spacecraft and shelters.
FRT	Flight readiness test.- The FRT is the launch pad test to establish that all spacecraft and launch systems are operating properly.
GSE	Ground support equipment.- The GSE includes the launch facilities and transporting, checkout, and servicing hardware.
ILMT	Integrated lunar module truck.- The ILMT is the LM descent stage (ascent stage removed) with attitude and guidance controls built into the structure to permit unmanned landings of logistics payloads.
IMU	Inertial measurement unit.- The IMU is the guidance equipment to determine the attitude reference.
LCG	Liquid coolant garment.- The LCG is thermal-control underwear worn inside soft or hard pressure suits.
LFU	Lunar flying unit.- The LFU is a short-range, one-man, thrusterpowered flying machine which can carry 300 pounds.
LM	Lunar module.- The LM is the standard Apollo two-man_ lunar landing spacecraft.
LMS	Lunar module shelter.- The LMS is the LM, stripped of ascent propulsion capability for crew habitation on the lunar surface.
LM&SS	Lunar mapping and survey system.- The LM&SS is a 4300-pound space module, containing high-resolution cameras, which is maneuvered by the docked CSM. The film is retrieved by an astronaut crawling through the docking tunnel.
LMTS	Lunar module truck shelter.- The LMTS is the LM descent stage with an inflatable human shelter and the capability to carry :heavy payloads.
LMT	Lunar module truck.- The LMT is the LM descent stage (ascent stage removed) modified to land unmanned with heavy logistic payloads.
LPM	Lunar payload module.- The LPM is the LM spacecraft stripped of all life support equipment, human accommodations and controls, and ascent propulsion capability to provide a minimum-modification unmanned lander.
LSSM	Local scientific survey module.- The LSSM is a short-range, 1000pound, one-man roving vehicle without a pressure cabin.
MCC-H	Mission Control Center.- The MCC is the manned space-flight tracking, communications, and remote command control center in Houston, Texas.
MFS	Manned flying system.- The MFS is a long-range, one- or two-man, thruster-powered lunar flying machine.
MLMT	Modular lunar module truck.- The MLMT is an unmodified LM descent stage (ascent stage removed) with docking, attitude, and guidance control modules added on top to provide an unmanned logistics payload lander.
MOBEX	Mobile excursion.- The MOBEX is a large, wheeled vehicle containing a pressurized cabin, a shelter, expendables, and power for long distance, long-duration lunar traverses.
MOLAB	Mobile laboratory.- The MOLAB is a large, wheeled vehicle containing a pressure cabin, a sheltered work space with scientific equipment, and expendables and. power for long-distance, long-duration lunar traverses: .
MSFN	Manned Space Flight Network.- The MSFN is a system of station complexes scattered around the earth for tracking, communicating with, and commanding the space missions.

MSS	<u>Mobil service structure</u>.- The MSS is a structure for working on the Saturn V and Apollo spacecraft at the launch pad.
OEM	<u>Orbital experiment module</u>.- The OEM is any of a number of modular instrument package concepts used to supplement the CSM lunar orbit remote-sensing capability.
PGA	<u>Pressure garment assembly</u>.- The PGA is a term for a soft EVA pressure suit.
PLSS	<u>Portable life support system</u>.- The PLSS is the EVA backpack for use with soft or hard pressure suits.
RCS	<u>Reaction control system</u>.- The RCS is the spacecraft attitude control thruster and propellant system.
RTG	<u>Radioisotope thermoelectric generator</u>.- The RTG is an electrical power and heat generator which uses radioactive fuels.
SFT	<u>Simulated flight test</u>.- The SFT is a test conducted on the launch pad during which all operational procedures are checked except booster ignition, to verify that the systems are functioning properly.
S-IVB	<u>Saturn IVB</u>.- The S-IVB is the third stage of the Saturn V launch vehicle.
SLA	<u>Spacecraft lunar module adapter</u>.- The SLA is the Saturn V top enclosure which supports the CSM and protects the LM derivatives, LM&SS payload, and so forth, during launch.
SM	<u>Service module</u>.- The SM is the Apollo spacecraft propulsion stage which also controls attitude, contains expendables and the EPS, and carries the science payload.
SPS	<u>Service propulsion system</u>.- 'The SPS is the main rocket engine and propellant tankage of the SM.
STEM	<u>Staytime extension module</u>.- The STEM is an inflatable human shelter and airlock for lunar surface use.
S-V	<u>Saturn V</u>.- The S-V is the three-stage Apollo launch vehicle.
TMG	<u>Thermal meteoroid garment</u>.- The TMG is the outer pressure suit covering for protection of the astronaut from solar heat, cold space radiation, and micrometeorites.

SECTION I

INTRODUCTION

The purpose of this publication is to provide the basic information and bare essentials required for lunar mission planning. Descriptions of the basic lunar space vehicles are provided, including the manned landers, the unmanned trucks, the mobility units, and the experiment packages. The payload capabilities of these vehicles are provided, along with operational constraints involved in the use of the vehicles. The intent is to keep the text simple and. easily understood by those unfamiliar with the engineering technology from which the vehicles were derived. The payloads obtainable with the various combinations of landed vehicles which might comprise a total mission; the weight requirements for the expendable items required for staytime; and the weight of flying and roving vehicles, plus their expendables, are included so that trade-off studies can be conducted between staytime, mobility, and scientific equipment. To avoid prejudicing those who must derive the missions, no attempt is made to describe actual missions that may be conducted on the lunar surface.

The scope of this report is the exploration time period beginning after the first two or three successful lunar landings of the Apollo Spacecraft Program and following through calendar year 1973. Limited information is provided for long-range roving vehicles or lunar bases that would be expected to be used in time periods beyond 1973. Only the basic facts for the various pieces of hardware are provided as they are best known at this date. The payload values and other data contained herein are considered to be reasonably accurate for mission planning purposes. Greater accuracy is not warranted in view of the fact that the basic weights of the spacecraft and the payload capabilities of the launch vehicles are continually changing, and the amount of change cannot be accurately predicted. Based on past space programs, weight; growth of the spacecraft has generally been compensated for by an increase in the performance of the launch vehicles. To keep this data book simple, no attempt is made to explain the reason the values are as listed or how they were derived unless it is essential to the understanding of the unit involved. References are provided where more detailed explanations and the sources for the information contained herein can be found. The vehicles discussed consist of the basic Apollo hardware and major modifications to the Apollo configurations, but no new major vehicle concepts are included. In addition to describing and providing data for the space vehicles with operational restrictions and constraints for the use of these vehicles at various locations on the lunar surface, an attempt has been made to provide the information on the capabilities and limitations of astronauts during extravehicular activities on the lunar surface. Also included is the astronaut time-line information for the time required for essential operational purposes and maintenance of health so that the total time available for experimental exploration can be derived. Cost information is provided to establish the effectiveness of the mission concepts and to give a feeling for the limitations imposed by the funding available for lunar exploration.

The information contained herein was derived from many sources. Some information was obtained from the National Aeronautics and Space Administration trajectory, performance, and cost analysis; other information resulted from funded studies; and still other data came from proposal literature and reports provided by various contractors. These data were verified to the extent possible, and the values were cross checked where available; however, these data may still contain some inaccuracies. The values presented are subject to change as further analyses are performed, as concepts more closely approach the tangible hardware stage, and as the spacecraft become operational and flight experience is gained.

SECTION II

SCHEDULES

Proper scheduling for the early lunar exploration program, as in all major programs, is extremely important. Few people realize the amount of time required from initial inception of a unit of hardware until the unit becomes available for actual launch. The time requirement applies to experimental equipment, spacecraft, launch vehicles, major pieces of supporting equipment, tools, instruments, and so forth. The schedule must include time for study and evaluation of the possible alternates involving the basic concept, the design time, the fabrication and development testing of the initial prototype, and sometimes redesign. Also, fabrication and assembly of flight-type articles, qualification testing of the units, modifications and changes dictated by test results, and assembly and checkout of the final flight units are required Simultaneously, the scheduling procedure requires the integration of the units into the associated vehicles or systems and into the training of the astronauts. The scheduled process takes approximately 3 to 5 years and is more dependent upon the staffing assigned to the task and the funding available than upon the size or complexity of the article. The total process can in some cases be shortened if extensive experience is available for similar hardware; if considerable related studies have been completed, the results of which are directly applicable to the new units; or if the new unit involves only moderate changes to previously built hardware.

The misconception which arises is that small instruments and experimental equipment can be built in small shops and placed on board the flight spacecraft in 2 years or less, and if the unit fails to produce results, not much is lost. Actually, the cost of sending a single pound of hardware payload to the lunar surface varies from $50 000 to $150 000 (vehicle costs divided by the landed payload). In addition, the weight which can be carried on the space vehicles is far too limited, and the time the astronaut is on the lunar surface is far too valuable to carry along any piece of equipment that does not have a very high degree of certainty for success before launch is attempted. Many excellent experiments of slightly lower priority are awaiting an opportunity to be flown. A number of experiments have been removed from past space flights because inadequate time was provided for development and qualification to cover contingencies in the tests, for integration into the flight system, or for properly training the astronaut in the use of the article.

Figure II-1 illustrates the earliest time in which various lunar spacecraft can be available for flight if implementation of the procurement process is started relatively soon. Delays add corresponding serial time to the availability dates. The delivery dates for the command and service modules, lunar modules, and Saturn V launch vehicles designated for lunar missions are presented in the charts following. These dates apply only to the present configuration. Modifications, reconfigurations, changes because of preceding flight experience, and reductions in funding all tend to slip the schedules. Also, delays in flight schedules will have profound effects. Some vehicles are presently slated for Apollo spares and others are unassigned. These vehicles could improve the dates shown on the charts if the vehicles were reassigned to early lunar missions. Such reassignments can also have major effects on the lunar program because it may be possible to complete most or all of the early lunar mission objectives without resorting to new-buy vehicles.

CSM

		Delivery date to KSC
117	Lunar mapping and survey system (LM&SS)	Nov. 1969
119	Spare Apollo landing	Mar. 1970
125	Lunar module (LM) shelter	Mar. 1971
127	LM taxi	July 1971
130	LM&SS	Dec. 1971
132	LM shelter	Mar. 1972
134	LM taxi	July 1972
141	LM shelter	Sept. 1973
142	LM taxi	Nov. 1973

LM

13	Spare Apollo landing	Oct. 1969
15	LM shelter	Feb. 1970
16	LM taxi	Jan. 1971
17	LM shelter	July 1971
18	LM taxi	Jan. 1972
20	LM shelter	Jan. 1973
21	LM taxi	June 1973

Saturn V

512	LM&SS	June 1969
513	Spare Apollo landing	Aug. 1969
516	LM shelter	June 1970
517	LM taxi	Dec. 1970
519	LM&SS	Dec. 1971
520	LM shelter	Mar. 1972
521	LM taxi	June 1972
525	LM shelter	June 1973
526	LM taxi	Sept. 1973

The chart depicts the shortest possible leadtimes for many major units of hardware for lunar exploration. These leadtimes also would be affected significantly by curtailment of funding.

Lunar unit	Shortest leadtime (from launch date), months
Extended lunar module (ELM) (3 day)	25
ELM (7 day)	28
Augmented lunar module (ALM)	42
Lunar payload module (LPM)	24
LM shelter	42
LM taxi	34
LM truck	33
ALM truck	42
Staytime extension module (STEM)	42
Early lunar shelter (ELS)	48
LM truck shelter	42
Lunar flying unit	36
Local scientific survey system (LSSS)	42
Mobile exploration, mobile laboratory	60
Drill (10 foot)	36
Drill (100 foot)	45
Staff	42

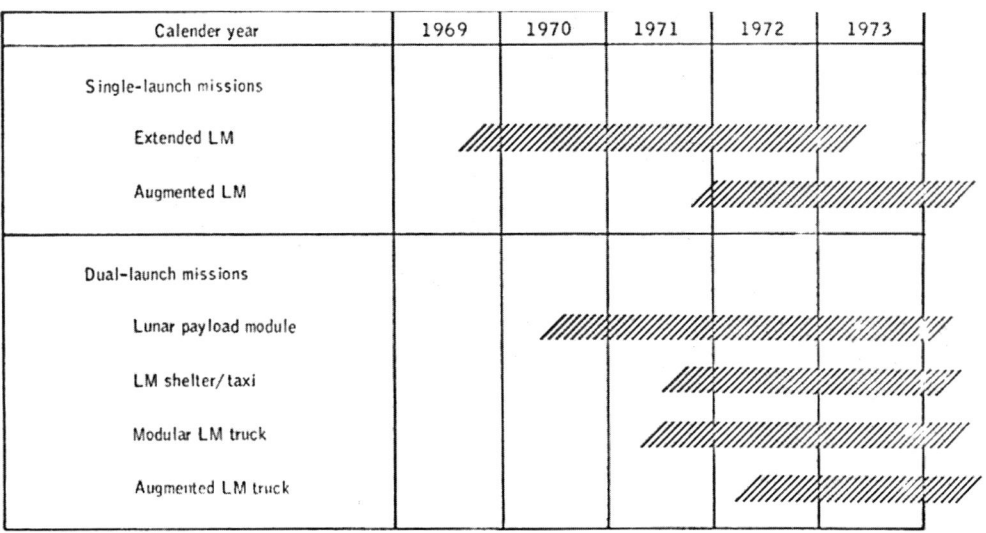

Figure II-1.- Availability and use periods of the lunar module.

SECTION III

MANNED LANDER DESCRIPTION*

Manned landers are vehicles used for the transport of two astronauts from the command module (CM) while it is in lunar orbit to the lunar surface and for subsequent return to lunar orbit for rendezvous with the CM. The manned landers described in this section are the Apollo lunar module (LM) and derivations of the vehicle.

Lunar Module

The LM is the lunar landing vehicle used in the Apollo program and is composed of two stages, the ascent stage and the descent stage (fig. III-1). The basic LM can deliver 300 pounds of scientific equipment to the lunar surface and can return 100 pounds of lunar samples, film, and containers to lunar orbit. Figure III-2 illustrates the overall dimensions for the LM.

Ascent stage.- The ascent stage (fig. III-3) houses the two-man crew, contains the automatic and manual equipment to control and monitor the vehicle and its system, and contains the ascent propulsion system for ascent from the lunar surface to lunar orbit. The crew compartment is pressure and temperature controlled and is used for both crew transport and living quarters during the lunar staytime. The usable floor area of the crew compartment is approximately 27 square feet, of which 14 square feet has head clearance and 13 square feet has only a 5-foot height. Total pressurized volume is approximately 250 cubic feet.

Descent stage.- The descent stage (fig. III-4) is the unmanned section of the LM and houses the descent propulsion system and extra oxygen and water tanks. The landing gear, used for landing on the lunar surface, is also attached to the descent stage. The descent stage is used as the launching pad for the ascent stage on return to lunar orbit.

*A list of definitions is included in this section.

Extended Lunar Module

The extended lunar module (ELM) extends the basic LM capabilities, with minimum modification of hardware, by deviating from the nominal Apollo ground rules. The deviations considered are (1) no free return, (2) lunar orbit separation at 50 000 feet versus 80 nautical miles, (3) range-free trajectory, (4) no continuous abort, and (5) optimum launch time. (Definitions of these terms are included at the end of this section.) All or part of the deviations determine the payload and surface-staytime extension capability up to 1500 pounds. The deviations can provide up to 7 days on the lunar surface, depending on the landing site location.

Surface staytime.- To enhance the surface staytime, system changes (the addition, of solar cells, radiators) will be required to utilize more efficiently the increased payload capability.

Additional payload.- Added expendables and scientific payload may be integrated within the structure where space is available. However, some of the added payload may have to be packaged and attached to the surface of the vehicle.

Hardware attachment.- Adequate volume is available for attaching hardware above the descent stage and around the ascent stage, but there are restrictions because of the requirement to be able to disconnect the LM stage and to be able to abort to lunar orbit at any time during the landing phase of flight.

Availability.- The ELM can be available early in 1970.

Augmented Lunar Module

The augmented lunar module (ALM) is a one-step redesign of the basic LM. The redesign attains a maximum landed payload of 2800 pounds, using a 98 000-pound-capacity Saturn V launch vehicle, to give a lunar staytime capability up to 14 days.

Redesign and modification.-.The redesign and modification is primarily to the descent stage. To land the heavier payload, a longer burn time is required from the present descent engine. To lengthen the burn time, the tank sizes are increased for greater propellant capacity, and ablative material is added to the engine expansion nozzle. Structural modification is necessary to accommodate both the increased system weight and the landing loads.

Availability.- Because of the design changes and the requalification requirements, the ALM could not become available for launch before 1972.

Definitions

Free-return trajectory.- A free-return trajectory would allow the spacecraft to circumnavigate the moon and return to earth without further propulsion from the spacecraft engines.

LM rescue.- The LM rescue is the capability of the CSM to descend from 80 nautical miles to 50 000 feet for active rendezvous with. the LM. This maneuver costs Delta-V from the CSM, however, the use of this technique can gain approximately 460 pounds in payload to the lunar surface.

Optimum launch times.- Launching at specific times requires less velocity because of the phasing between the earth and the moon. This difference in velocity requires less propellant which may be exchanged for increased payload.

Trajectory time.- Trajectory time is the time required to travel from the earth to the moon, and vice versa. The shorter the time, the higher the Delta-V requirements resulting in more propellants being used.

Continuous abort capability.- This is the capability of aborting from the lunar surface at any time for rendezvous with the CSM. The LM has a 0.5° plane change capability. Orbital precision increases with the higher latitudes and orbit inclinations and tends to make continuous abort impractical.

Range-free trajectory.- Range-free trajectory is a trajectory in which range is not constrained and in which guidance equations for steering allow for range errors at touchdown. The greater the allowable error is, the lower the fuel consumption for correction will be.

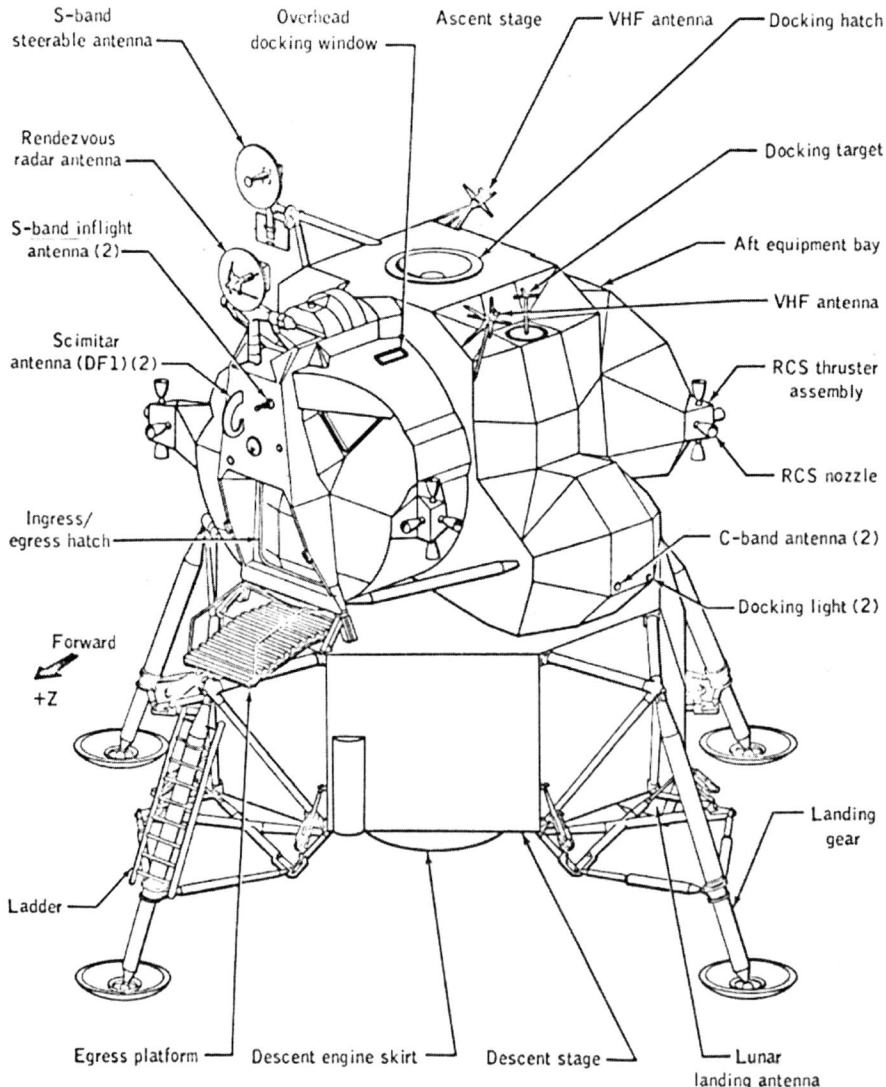

Figure III-1.- Lunar module configuration.

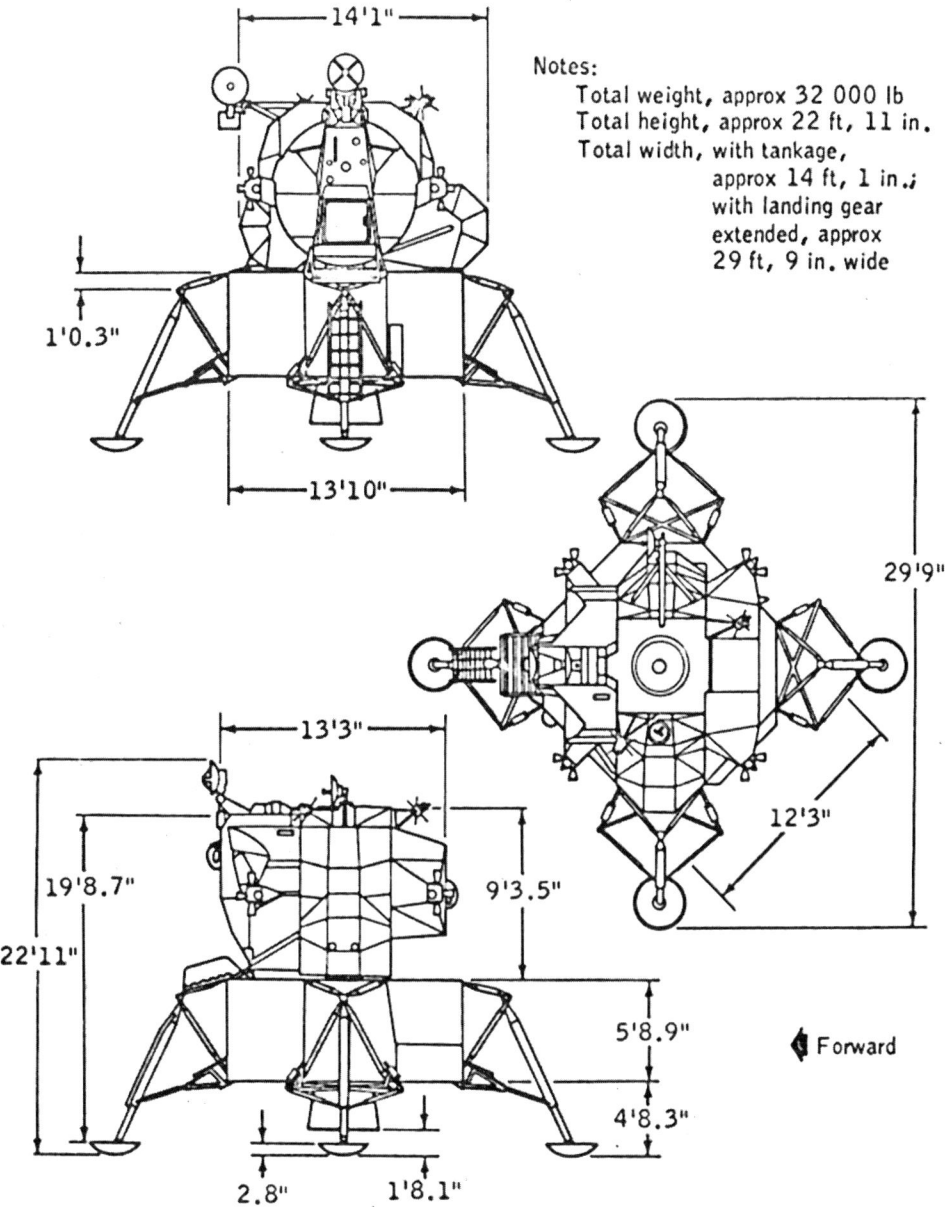

Figure III-2.- Lunar module overall dimensions.

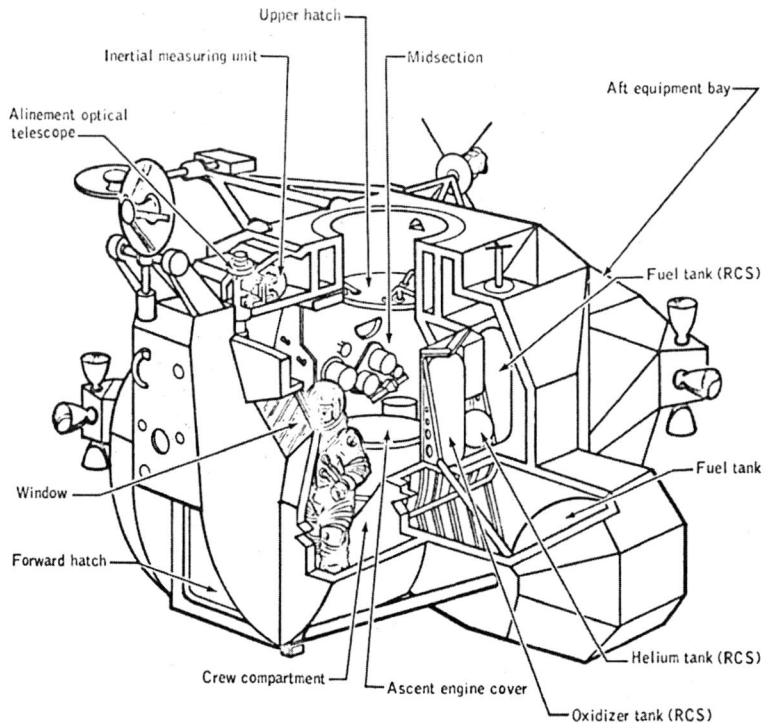

Figure III-3.- Lunar module ascent stage.

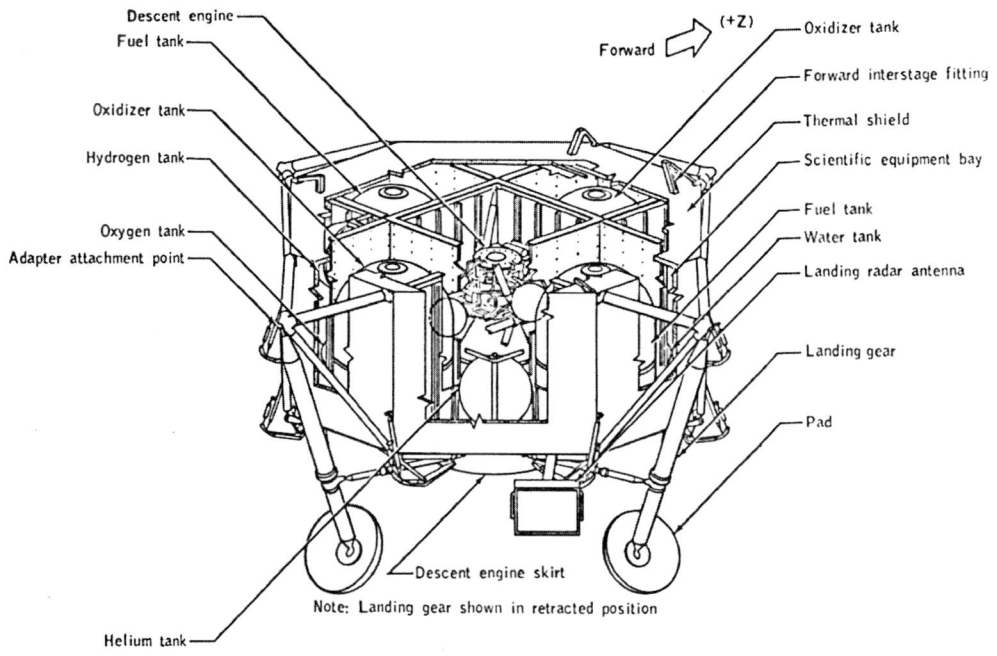

Figure III-4.- Lunar module descent stage.

SECTION IV

UNMANNED LANDER DESCRIPTION

Unmanned landers are derived from the lunar module (LM) vehicle and could be delivered to lunar orbit attached to a manned command module (CM) and/or a service module (SM). Payload weight and volume are increased considerably with an unmanned lander. Elimination of the abort restrictions removes many of the payload constraints and makes available a considerable volume above the descent stage. Landing impact is less critical because the vehicle is unmanned and does not have an ascent capability. However, because the vehicle is unmanned, there is no obstacle-avoidance capability. Landing accuracy is expected to be comparable to the Apollo LM. Experience with the present hardware and with new techniques means that accuracies within 1000 feet are possible with the unmanned vehicle.

Lunar Payload Module

The lunar payload module (LPM), sometimes referred to as the stripped LM, is basically an LM with the main ascent propulsion system, the life support system, and the crew controls removed. The vehicle is modified to land a payload on the lunar surface unmanned and to sustain the payload in a quiescent state for, up to 3 months. The LPM payload capability is approximately 8200 pounds with an available volume of approximately 730 cubic feet. Basing the prediction on a go ahead for phase C in the first quarter of 1968, the earliest launch date for the LPM would be the first quarter of 1970.

Augmented lunar payload module.- An augmented LPM would be similar to the basic LM augmentation. The payload would be increased to 10200 pounds and have essentially the same cargo volume capability.

Lunar Module Truck

The LM truck is an LM descent stage which can land an unmanned payload on the lunar surface. Two basic truck concepts or philosophies are proposed (fig. IV-1) and are shown as the integrated LM truck and the modular LM truck.

Integrated lunar module truck.- The integrated LM truck has the guidance and control systems integrated within the descent structure, with some structural modification and requalification of the descent stage anticipated. The payload capability is approximately 10000 pounds with a payload volume of approximately 2000 cubic feet.

Modular lunar module truck.- The modular LM truck has a module incorporating the guidance and control systems. This module is added to the top of the descent stage and requires essentially no modification to the descent stage. The payload volume is almost the same as the integrated LM truck, but the payload weight is reduced approximately 200 pounds because of the module structure weight.

Lunar module truck docking adapter.- The LM truck has the same need for a docking adapter as does the LM for lunar orbit delivery. Three concepts are considered for attaching the docking adapter: (1) the pylon or center post method, (2) the bridge method, and (3) the method of mounting a docking mechanism directly onto the payload package or container. The advantage of each concept depends upon the type of payload to be carried.

Augmented lunar module truck.- The augmented LM truck would be similar to the augmented lunar module (ALM) described in section III. The payload capacity would be increased to approximately 12 000 pounds.

Direct Delivery of Unmanned Landers

The Surveyor demonstrated the direct delivery of an unmanned vehicle to a soft landing. The modular LM truck could make direct delivery by the addition of translunar guidance and command capability to LM truck guidance equipment presently planned.

Useful payload.- The payload landed on the lunar surface is limited by the capacity of the LM descent stage. However, the direct delivery method would provide the capability for leaving a large payload in lunar orbit.

Vehicle.- To take advantage of the payload capability of the direct delivery approach, a completely new lander is required, together with one or two new launch vehicle stages. A vehicle of this type could be made available in 1975, should the need be justified.

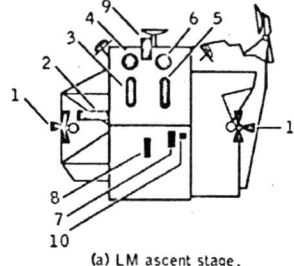

Existing components
1. RCS thrusters
2. Battery
3. RCS oxidizer
4. RCS helium
5. RCS fuel
6. ECS water
7. ECS evaporator
8. Guidance computer
9. IMU star tracker
10. S-band ERA

(a) LM ascent stage.

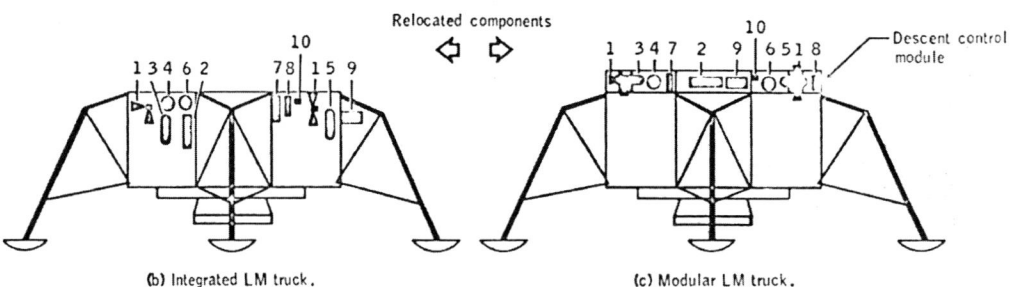

(b) Integrated LM truck.

(c) Modular LM truck.

Figure IV-1.- Lunar module truck concepts.

SECTION V

LUNAR ORBIT MISSION CAPABILITY

During lunar mapping survey flights and lunar payload deliveries, either manned or unmanned, opportunities exist to include instrumentation on board the command and service module (CSM) to measure lunar parameters from lunar orbit.

Lunar Mapping and Survey System

On the lunar mapping and survey system (LM&SS) flights, which are conducted at an altitude of approximately 80 nautical miles, the lunar mapping and survey system module is carried in the space usually occupied by the lunar module (LM). Spaces which are available to carry instruments include bay 1 of the service module (SM), the space between the command module (CM) and the SM, and possibly space in the cabin. Bay 1 has a total volume of approximately 170 cubic feet, a length of 13 feet, and from the astronauts' position is located just to the right of the hatch (fig. V-1). The space between the CM and the SM is relatively narrow (12 to 15 inches), but space is available in several areas around the circumference between the radiators. Location of instruments outside the cabin might require extravehicular activity (EVA) for the retrieval of film, strip charts, and similar items of equipment. Bay 1 of the SM could be used for mounting instruments for lunar orbit measurements, or the bay could be used for unmanned probes or subsatellites such as the anchored interplanetary monitoring platform (AIMP) (fig. V-2). For example, on a mapping and survey system flight using a 98 000-pound-boost Saturn V for a 14-day lunar polar orbit and a non-return of the instruments from lunar orbit to earth orbit, an average of 11 000 pounds of instruments could be carried with the present CSM dry weight of 24 000 pounds. The total instrument payload would vary depending on trajectory, launch windows, time of launch, and final CSM gross weights. If the instrumentation payload is to be returned to earth orbit because of its installation or location, the instrumentation payload weight would be reduced to two-thirds of the original weight. The following are typical instrument payloads:

	80-nautical-mile orbit		50 000-ft orbit[a]	
	Example 1, lb	Example 2, lb	Example 1, lb	Example 2, lb
Unmanned delivery				
Lunar orbit payload	3000	2000	1500	1000
Lunar surface payload	8000	8000	8000	8000
Earth-returned payload[b]	250	2000	250	1000
Manned delivery				
Lunar orbit payload	3000[c]	2000[c]	100[d]	2200[d]
Lunar surface payload	300[c]	300[c]	1500[d]	300[d]
Earth-returned payload from lunar orbit[b]	100[c]	2000[c]	0[d]	2200[d]
Lunar surface to earth payload	150[c]	150[c]	150[d]	150[d]

a Pickup, 40 nautical miles.
b The CM has a volume sufficient for only about 250 pounds of payload through reentry and earth landing. All additional payload is separated prior to reentry.
c LM active rendezvous.
d CSM active rendezvous.

The manned delivery at 50 000 feet uses the LM weight of 29700 pounds. All other lunar modules use 32600 pounds. See section VII.

Manned and Unmanned Lunar Landings

The lunar-orbital flights required to deliver a manned or unmanned spacecraft to the lunar surface are effective tools for performing other useful orbital activities. These other activities include lunar-orbital instrumentation measurements and the launching of subsatellites or probes. Some of the lunar orbit payload could be expended in providing the ability to change the lunar orbit for more effective use of cameras or remote sensors. This would allow the orbiting astronauts to pick or select sites of opportunity for photographs and remote sensing.

Figure V-3 shows an orbital experiment module (OEM) which could be placed above the LM spacecraft or LM truck during ascent and later could be attached to the small end of the CM. The OEM could give access to the tunnel section of the CM, allowing retrieval of film, strip charts, and similar items of equipment. Figure V-4 shows the location of the OEM during lunar orbit. The OEM could be jettisoned in lunar orbit after completion of all measurements and retrieval of the data. Other locations noted in the mapping and survey system flights are available as potential instrument carriers.

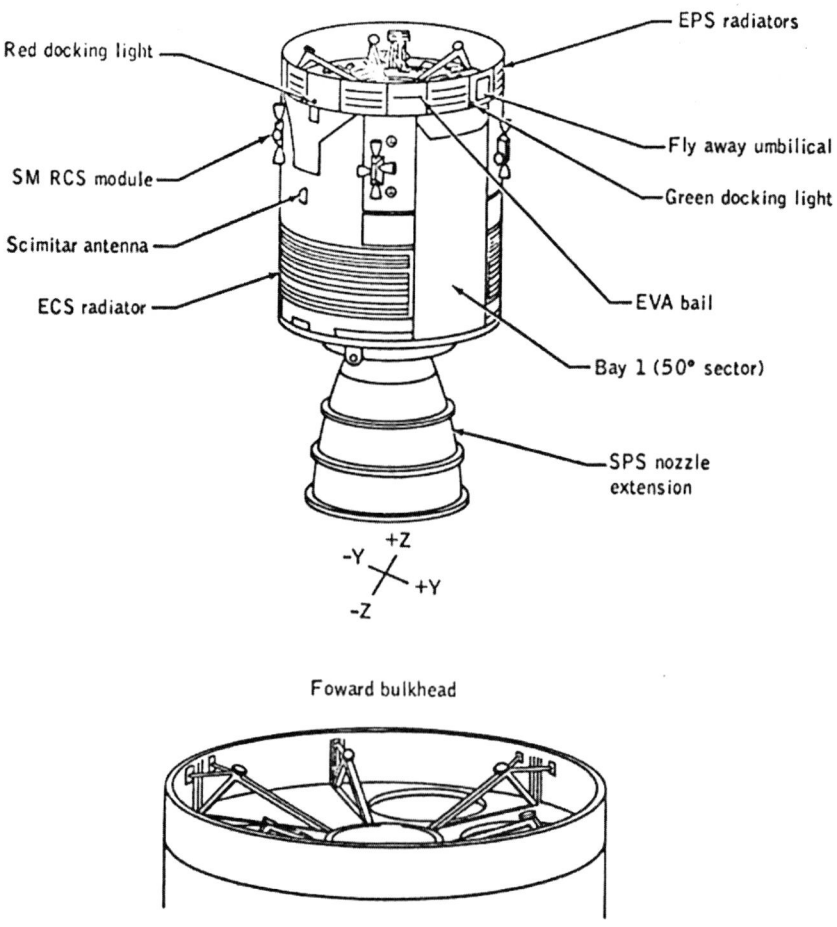

Figure V-1.- Service module, block II.

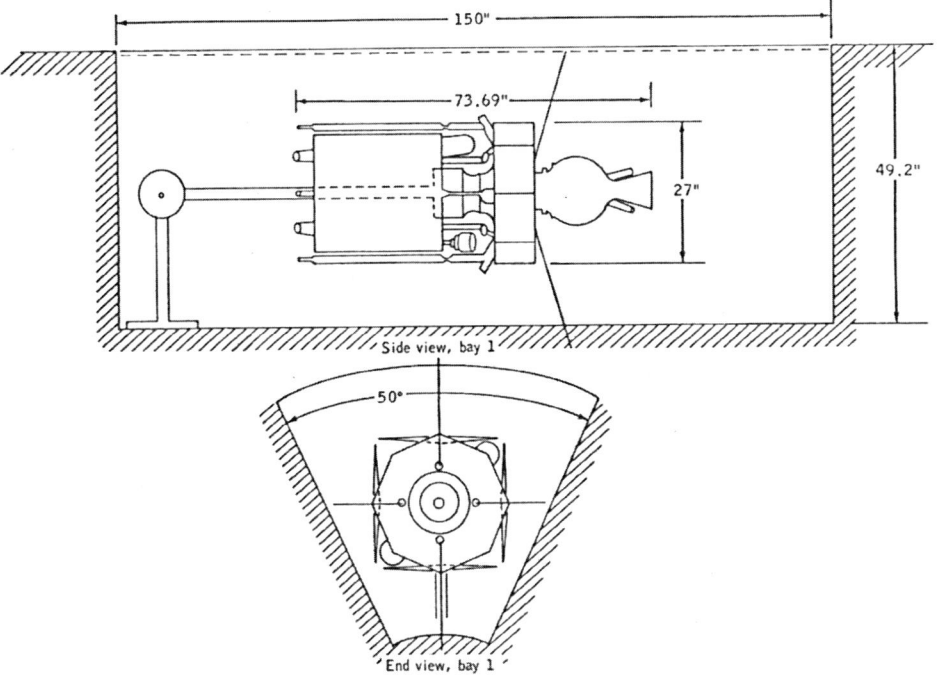

Figure V-2.- The anchored interplanetary monitoring platform.

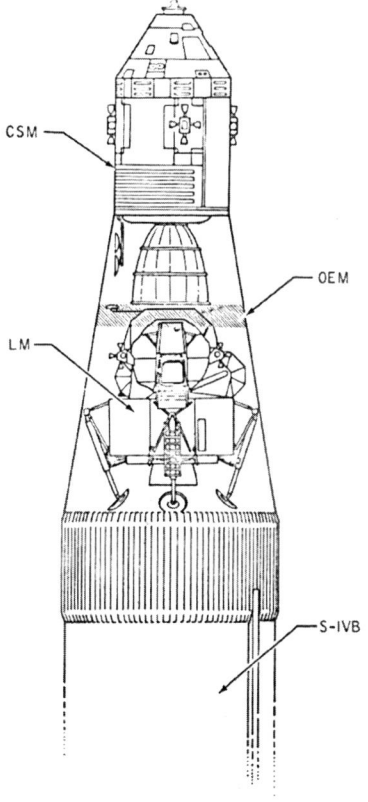

Figure V-3.- Orbital experiment module, launch configuration.

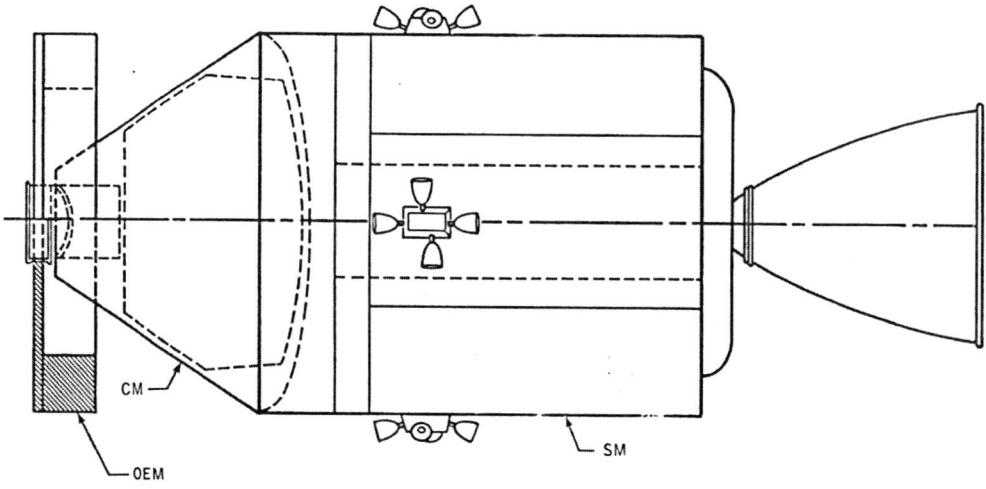

Figure V-4.- Orbital experiment module, lunar orbit configuration.

SECTION -VI

AUXILIARY SHELTER AND TRUCK SHELTER VEHICLES

Lunar Module Shelter

System description.- The lunar module shelter (LMS) (fig. VI-1), is an Apollo lunar module (LM) with modifications for an unmanned landing on the lunar surface. The shelter remains in a stored condition for a period of up to 90 days and then supports two men for up to 14 days for exploration and scientific missions.

Before occupancy and prior to launching the astronauts, the shelter can be checked out from earth. The shelter can be activated by remote control either from the earth or from the manned lander after a successful landing. Final activation and checkout are performed by the astronauts. The shelter has a gross volume of 340 cubic feet, which includes 252 cubic feet for the cabin and 88 cubic feet for the airlock. Work areas are provided for scientific experimentation (fig. VI-2). Bunks permit the astronauts to sleep in a semi-reclining position (figs. VI-3 and VI-4).

Major system modifications and/or changes from the standard LM are the removal of the ascent stage engine; conversion of the electrical power system from all batteries to a combination of batteries and fuel cells; provision for cryogenic oxygen and hydrogen storage; the addition of radiators for cooling; the addition of crew provisions, work areas, and bunks; and the possible addition of an airlock.

General performance.- General performance characteristics of the shelter are as follows:

1. Electrical power of 94 kWh available for-scientific activity, 28-volt nominal with 3-kW peak rate

2. A 5275-pound scientific and mobility payload for the standard descent stage, exclusive of staytime expendables

3. A 7275-pound payload with an augmented descent stage

4. An available payload area of 471 cubic feet

5. An additional payload area of 100 pounds available for each day of staytime reduction below 14 days.

Development time.- The expected development time is approximately 3 years.

Development cost.- The development cost is $119 million. Unit cost.- The unit cost is $42 million.

Early Lunar Shelter

System description.- The early lunar shelter (ELS) is a rigid structure horizontal cylinder with a gross volume of 750 cubic feet, mounted on an LM truck. The shelter is designed for an unmanned landing on the lunar surface where it, can remain stored for up to 180 days prior to supporting two men for up to 50 days of exploration and scientific activity.

The shelter can be checked out from earth prior to launching; the vehicle which carries the crew for the ELS. Initial activation of the shelter is either from the earth or from the manned lander after a successful landing. Final activation and checkout are performed by the astronauts.

The configuration of the ELS is shown in figure VI-5. The airlock supports two men at one time with ample work and sleep areas provided in the cabin (fig. VI-6).

The electrical power consists of batteries and a SNAP-27. radioisotope thermoelectric generator (RTG) for storage. Batteries and fuel cells provide power during occupancy. Hydrogen and cryogenic oxygen storage is utilized with fuel cell product water for food reconstitution and drinking and for cooling with the water sublimators. Cooling is accomplished with a water/glycol transfer loop, sublimators, and radiators. Storage areas are provided within the pressurized cabin for hard suits, food, lithium hydroxide (LiOH) canisters, and certain experiments. However, larger equipment for experiments and items such as a local scientific survey module (LSSM) or a flying vehicle must be mounted in the restricted area exterior to the ELS or in the landing gear strut area (fig. VI-1). The total storage capacity of these areas is approximately 367 cubic feet.

General performance.- General performance characteristics of the ELS are as follows:

1. Electrical power of 600 kWh available for scientific activity, 28-volt nominal with 3.5-kW peak rate

2. A remaining 3200-pound scientific and mobility payload with a standard LM truck used for descent

3. A 5200-pound payload with an augmented truck used for descent

4. An additional payload of 54 pounds available for each day of staytime reduction below 50 days

Development cost.- The development cost is $165 million.

Unit cost.- The unit cost, including the cost of the LM truck, is $62 million.

Staytime Extension Module

System description.- The staytime extension module (STEM) (fig. VI-7) is an expandable structure capable of supporting two men on the lunar surface for 8 to 14 days. The module can be delivered in canisters attached to the side panels of the augmented lunar module (ALM) descent stage or attached to the LM truck. The structure is placed on a thermal mat on the lunar surface. After it is unpackaged, the structure becomes self-erecting. Equipment modules are placed within the shelter after the shelter is unloaded from the descent stage. The total volume of the shelter is 515 cubic feet of which 410 cubic feet form the cabin area and 105 cubic feet comprise the airlock. The total mass of shelter, systems, and consumables for an 8-day mission is approximately 1600 pounds. Power is provided by fuel cells with cryogenic oxygen and hydrogen storage. However, power can be provided by solar cells for an all-day mission. Communications consists of four basic links as follows: shelter to CM, shelter to LM, shelter to exploring astronauts, and shelter to earth.

The STEM provides ample work and rest areas within the cabin and a two-man airlock with access at the lunar surface level. Additional development work is necessary to insure that the ALM control thruster plume impingement on the packaging containers is not a problem, that the thermal control system is adequate, and that no fire hazard exists because of the flexible structure.

General performance.- General performance characteristics of the STEM are as follows:

1. Although electrical power is not provided for scientific activity with the fuel cell power system, the fuel cell power system can be either increased in size or replaced by a solar array sized to provide the power for scientific activity.

2. A payload of 1200 pounds of scientific and mobility equipment is available with the AIM in addition to the mass of the STEM.

3. Approximately 67 cubic feet is available between the landing gear support struts for the payload (fig. VII-1, item D).

4. Approximately 60 pounds additional payload is available for each day of staytime reduction below 8 days for the fuel cell configuration.

5. Approximately 60 pounds of additional expendables plus tankage is required for each day of staytime beyond 8 days for the fuel cell configuration.

Development time.- The development time is approximately 3-1/2 years.

Development cost.- The development cost is $46 million.

Unit cost.- The unit cost is $8 million.

Lunar Module Truck Shelter

System description.- The lunar module truck shelter (LMTS) (fig. VI-8) is an LM truck or modular LM truck with a rigidly attached docking bridge. The shelter contains a rigid airlock, shelter controls, consumables, and systems. Attached to the docking bridge are the flexible walled cabin areas which collapse to increase cargo volume during delivery of the shelter to the lunar surface. The airlock on the bridge can be used for rest prior to cargo unloading; and total activation of the shelter. The LMTS is delivered to the lunar surface unmanned. The

Auxiliary Shelters & Truck Shelter Vehicles

shelter is checked out in a manner similar to that described for the early lunar shelter.

Development.- The development cost is $107 million, and the time is approximately 3 to 4 years.

Unit cost.- The unit cost is $39.5 million.

Lunar Shelter Comparison

Table VI-1 provides a comparison of capabilities of various lunar shelters.

TABLE VI-I.- LUNAR SHELTER COMPARISONS[a]

Module	Lunar staytime, days	Scientific payload Mass, lb	Scientific payload Volume, cu ft	Cabin volume, cu ft	Airlock volume, cu ft	Cabin floor area, ft	Dry mass, lb	Expendables, lb	Figure
Lunar module shelter	14	5275 [b]7275	471	252	88	27	3325	1400	VI-1 VI-2 VI-3 VI-4
Early lunar shelter	50	3200 [b]5200	367	628	122	68	4100	2700	VI-5 VI-6 VI-7
Staytime extension module[c]	8	1200	67	410	105	52	1060	540	VI-7
Lunar module truck shelters	14	5800 [b]7800	1300	550	120	110	2800	1400	VI-8

[a]Mass, volume, and so forth, are those which are available on the shelter transport vehicle only and do not include payload available on taxi vehicles for dual-launch missions.

[b]Indicates mass payload with augmented descent stage.

[c]Figures given for STEM based on single-launch mission with ALM. All other shelters require dual launches.

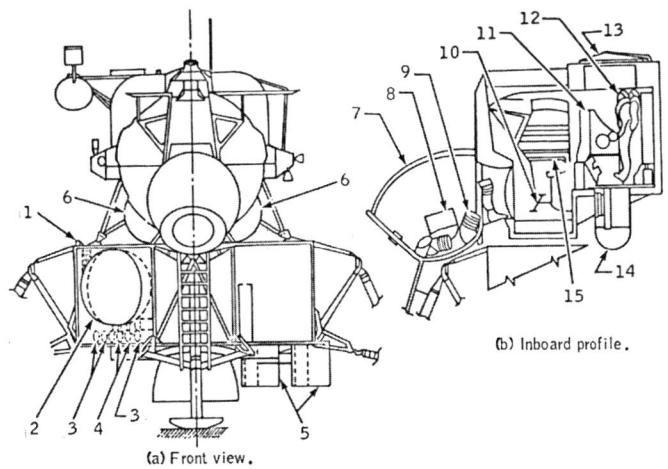

(a) Front view.　　(b) Inboard profile.

1. RTG
2. Oxygen tank
3. Cabin LiOH canister
4. PLSS LiOH canister
5. PLSS
6. Hydrogen tanks
7. Airlock (expanded)
8. PLSS
9. Hard suit (stowed)
10. Exerciser
11. Environmental control system
12. Soft suits (stowed)
13. Upper thermal hatch cover
14. Waste management system
15. Work station

Figure VI-1. Lunar module shelter.

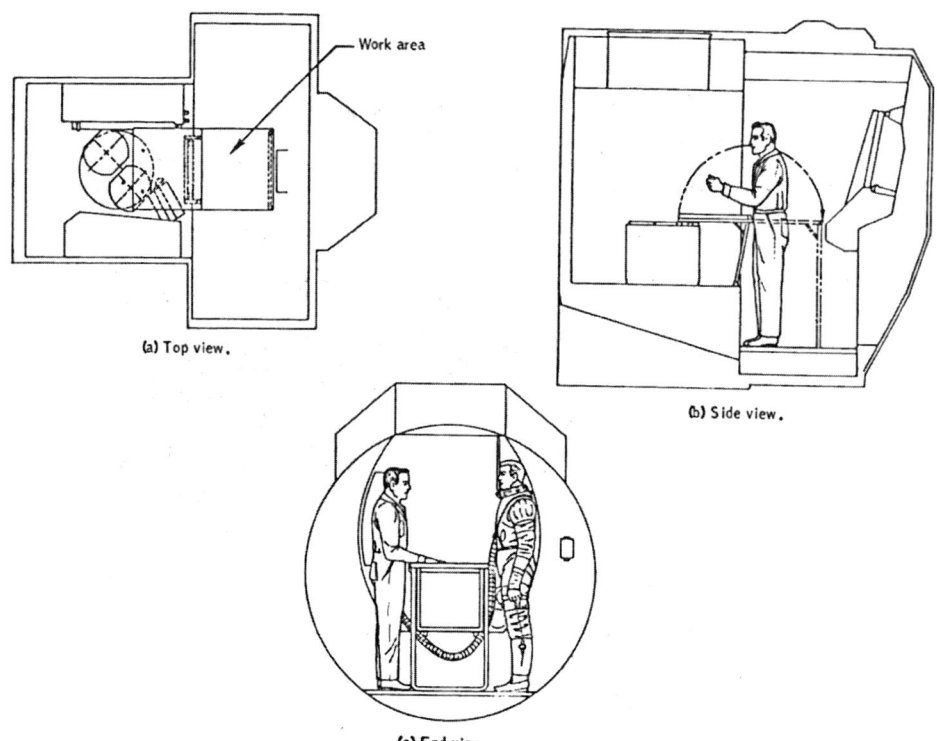

Figure VI-2.- Lunar module shelter work area.

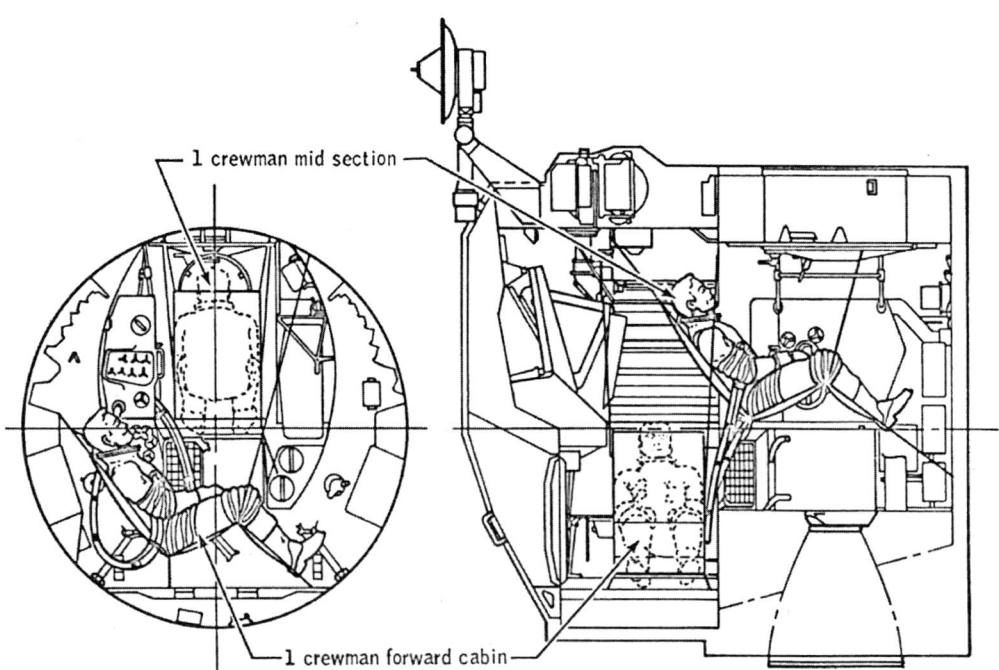

Figure VI-3.- Sleeping arrangement concept.

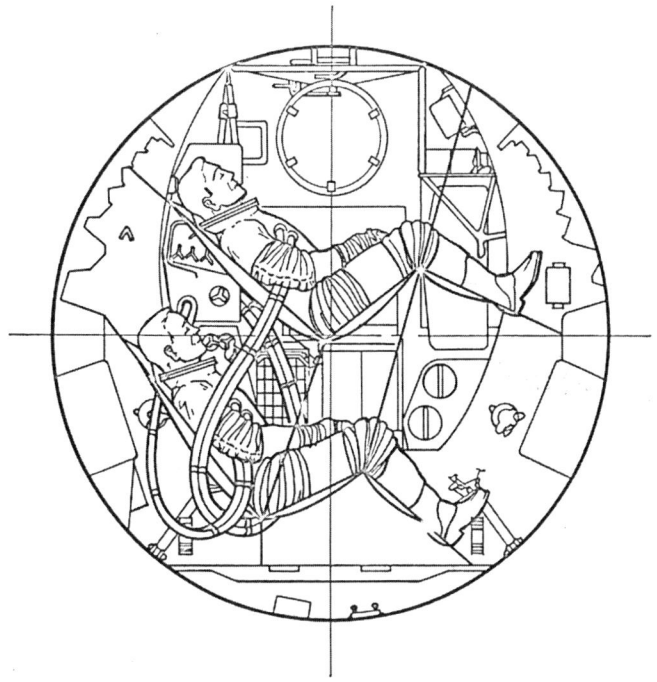

Figure VI-4.- Sleeping arrangement concept, forward cabin.

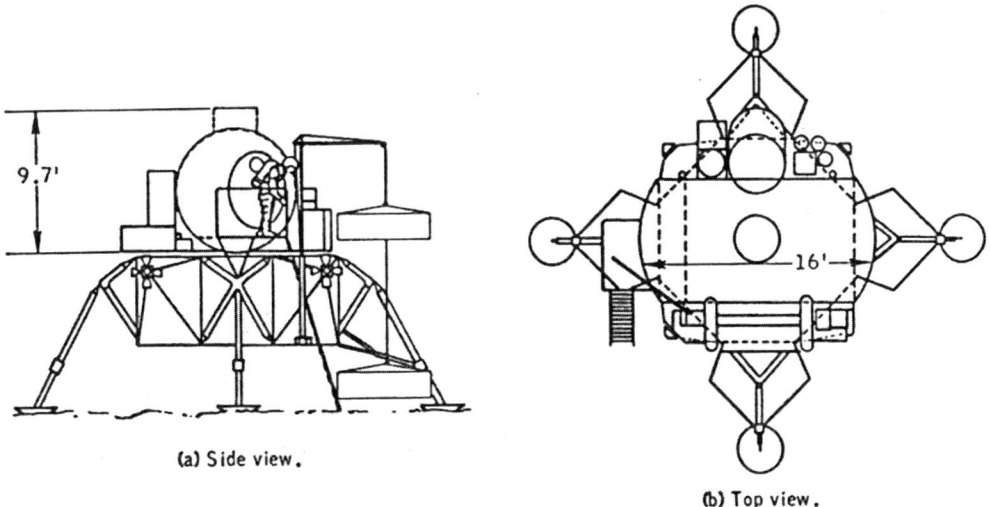

(a) Side view.

(b) Top view.

Figure VI-5.- Early lunar shelter.

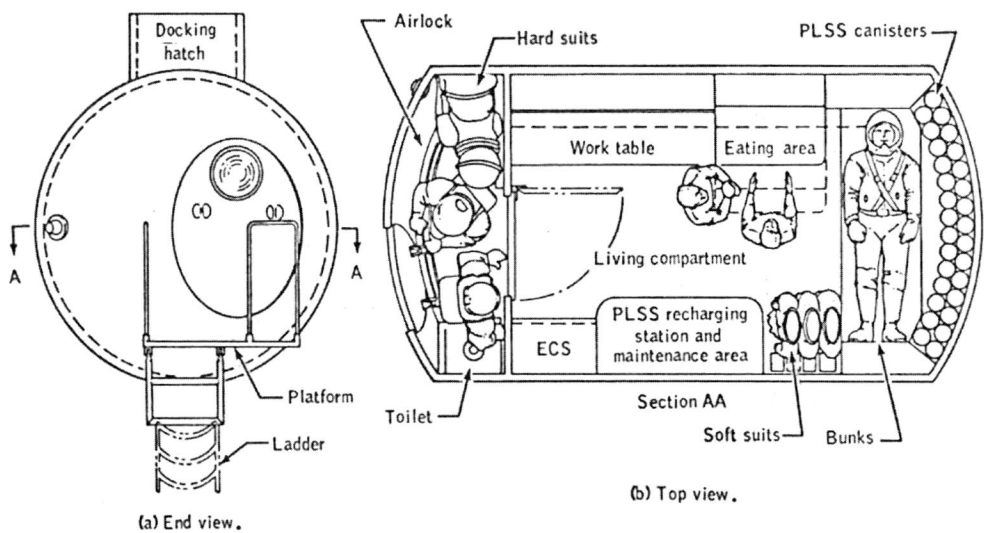

Figure VI-6.- Early lunar shelter, detail view.

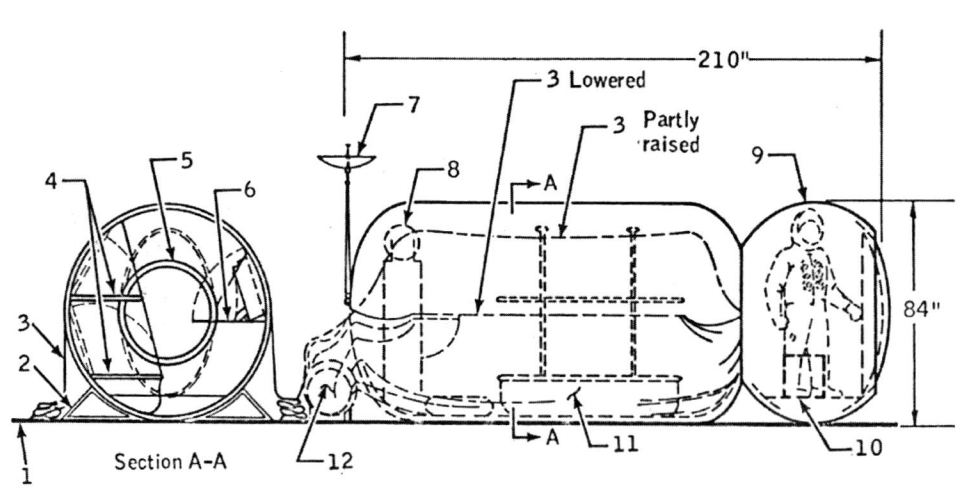

1. Thermal mat
2. Portable chocks (4)
3. Thermal control blanket
4. Bunks
5. Airlock inner hatch 36-in. diam.
6. Backpack storage and worktable
7. Antenna
8. Water
9. Airlock
10. Toilet
11. Spacesuit storage
12. Cryogenics storage

Figure VI-7.- Staytime extension module.

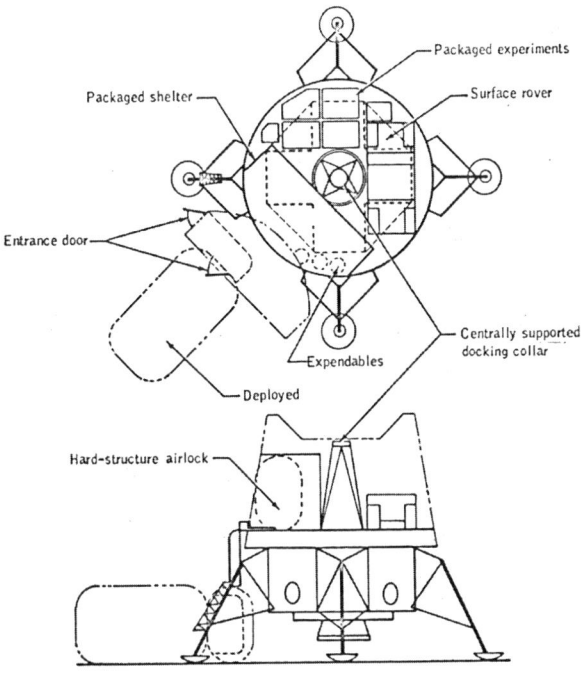

Figure VI-8.- Lunar module truck shelter.

SECTION VII

PAYLOAD CAPABILITIES

This section presents a brief summary of the best estimates of the lunar-surface payload delivery capability of several spacecraft and spacecraft combinations. An indication of the payload delivered to lunar orbit is also presented.

Surface Payload

Table VII-I presents the total landed payload (including expendables), the required expendable weights (as a function of surface staytime), and combinations of experimental scientific payload and range capabilities when two selected lunar flying vehicles (400 pounds including ground support equipment (GSE)) are used. See section X for flying vehicle descriptions. Table VII-I indicates the total range achievable for a flying vehicle as a function of the experiment package weight and surface staytime.

An equatorial landing site is assumed for these surface payloads. The total landed payload at a site off the lunar equator is dependent upon the site longitude and latitude. No data are presented for non-equatorial landings. In general, payloads for non-equatorial landings are less than those indicated in the tables.

Consideration should be given to the volume available for the pay-load packages. A very limited internal space is available in the manned landing vehicles. External mounting can be employed on the extended lunar module (ELM), but the indicated payload weight includes the structural modifications necessary to carry the additional payload. Little additional external space will be found on the augmented lunar module (ALM) for payload-package mounting.

The additional volume available in the lunar payload modules (LPM) vehicle is that space normally available for the crew and that space obtained by removing the ascent propulsion system. The LM truck: obtains additional space by replacing the LM ascent stage with a payload package.

The LM taxi and shelter vehicles are again volume restricted because both LM ascent stages must provide crew space. Figure VII-1 illustrates a method for packaging the surface payloads. Packaging of this nature will, in general, require LM structural modifications. If large center-of-gravity changes result, the LM guidance-and-control and stabilization-and-control systems will have to be investigated and possibly modified.

Orbital Payloads

Table VII-I shows data for missions that enhance surface payload, and as such, may result in no significant scientific payload in lunar orbit. The lunar orbit payload available is dependent upon the landing site and may be as heavy as 3000 pounds for some single manned missions; however, most landing missions will not; have significant lunar orbit payload capability unless the mission is specifically designed for that purpose.

Missions designed for the purpose can deliver large payloads to lunar orbit if no landing is made. Table VII-II indicates the orbital payloads that are attainable for some selected missions. Polar orbits are employed to obtain the maximum surface coverage during the lunar orbit staytime. All indicated payload is left in lunar orbit and cannot be returned to earth. Each 2 pounds of payload returned to earth reduces the indicated values by approximately 3 pounds.

Flights 1 and 2, listed in table VII-II, are required to over-fly the western and eastern quadrants (near-earth side), respectively. The 14-day command and service module (CSM) lifetime that is used does penalize the deliverable payload.

Flight 3 in table VII-II indicates an orbital payload capability of greater than 30 000 pounds. This flight employs a CSM modified for 36-day operation which spends 28 days in lunar polar orbit to obtain complete coverage of the lunar surface. The 28 days in orbit are required if specific surface lighting is desired.

Because of the large payloads obtainable for the flights, no effort is made to establish precise values. The payloads shown in table VII-II are the minimums that are always available when allowances are made for mission planning flexibility.

The volume available for the lunar-orbital mission is that space normally allotted for the LM and possibly some small volume in the CSM. Much larger volumes are available for payloads on orbital flights than on landing flights.

Payload Capabilities

TABLE VII-I.- LUNAR SURFACE PAYLOADS

Vehicle	Payload, lb	Staytime, days	Expendables,[a] lb	Experiments, lb	Mobility aid	Range, miles
Standard LM	250	1.5	---	250	---	---
ELM	1500	2.0	---	250	F[b]	120
		5.0	850	250	---	---
		7.0	1100	250	---	---
ALM	2800	3.0	550	250	F	170
		7.0	1100	250	F	130
		10.0	1500	250	F	110
		14.0	2050	700	---	---
ELM, 2 (1 manned + 1 unmanned) 1500 lb + 2100 lb	3600	3.0	550	250	F	160
		5.4	880	250	F	270
		7.0	1100	250	F	250
		3.0	[c]1100	250	F	120
		7.0	[c]2200	200	F	180
ALM, 2 (1 manned + 1 unmanned) 2800 lb + 3400 lb	6200	7.0	1100	250	F	420
		7.0	1100	700	F	390
		7.0	[c]2200	700	F	320
		10.0	1500	700	F	370
		10.0	[c]3000	700	F	260
		14.0	2050	700	F	330
		14.0	[c]4100	700	F	190

[a]Includes actual consumables plus tankage, additional batteries, solar panels, radiators, and so forth, as may be appropriate to extend staytime to the indicated value.

[b]F = flyers.

[c]Full expendables for both vehicles.

TABLE VII-I.- LUNAR SURFACE PAYLOADS - Continued

Vehicle	Payload, lb	Staytime, days	Expendables,[a] lb	Experiments, lb	Mobility aid	Range, miles
ELM + LPM, 1500 lb + 8400 lb	9 900	7.0	1100	700	FR[b]	550
		10.0	1500	700	FR	530
		14.0	2050	700	FR	490
ALM + LPM, 2800 lb + 8400 lb	11 200	7.0	1100	700	FR	640
		10.0	1500	700	FR	610
		14.0	2050	700	FR	580
ELM + LM truck, 1500 lb + 10 000 lb	11 500	7.0	1100	700	FR	660
		10.0	1500	700	FR	640
		14.0	2050	700	FR	600
ALM + LM truck, 2800 lb + 10 000 lb	12 800	7.0	1100	700	FR	750
		10.0	1500	700	FR	730
		14.0	2050	700	FR	690
ALM + augmented LPM, 2800 lb + 10 400 lb	13 200	7.0	1100	700	FR	770
		10.0	1500	700	FR	760
		14.0	2050	700	FR	720

[a]Includes actual consumables plus tankage, additional batteries, solar panels, radiators, and so forth, as may be appropriate to extend staytime to the indicated value.

[b]FR = flyers and rovers (LSSM). The rover contributes 5 to 10 percent of the indicated range with maximum rover range of 120 nautical miles.

TABLE VII-I.- LUNAR SURFACE PAYLOADS - Concluded

Vehicle	Payload, lb	Staytime, days	Expendables,[a] lb	Experiments, lb	Mobility aid	Range, miles
ALM + ALM truck 2800 lb + 12 000 lb	14 800	7.0 10.0 14.0	1100 1500 2050	700 700 700	FR[b] FR FR	880 860 820
ELM taxi + LM shelter 1500 lb + 5000 lb	6 500	7.0 10.0	[c]2200 [c]3000	700 700	FR FR	260 200
ALM taxi + ALM shelter 2800 lb + 7000 lb	9 800	7.0 10.0 14.0	[c]2200 [c]3000 [c]4100	700 700 700	FR FR FR	480 420 350
ELM taxi + LMTS 1500 lb + 8000 lb	9 500	7.0 10.0 14.0	[c]2200 [c]3000 [c]4100	700 700 700	FR FR FR	460 400 330
ALM taxi + augmented LMTS 2800 lb + 10 000 lb	12 800	7.0 10.0 14.0	[c]2200 [c]3000 [c]4100	700 700 700	FR FR FR	680 620 550

[a] Includes actual consumables plus tankage, additional batteries, solar panels, radiators, and so forth, as may be appropriate to extend staytime to the indicated value.

[b] FR = flyers and rovers (LSSM). The rover contributes 5 to 10 percent of the indicated range with maximum rover range of 120 nautical miles.

[c] Full expendables for both vehicles.

TABLE VII-II.- LUNAR-ORBITAL MISSION PAYLOAD

[Polar orbits, payload abandoned in lunar orbit]

	Mission duration, days	Payload, lb	Comments
Flight 1	14	18 800	Overflies 0° to 90° W quadrant
Flight 2	14	16 700	Overflies 90° E to 0° quadrant
Flight 3	36	30 000	Requires extended CSM, 28 days in lunar orbit, overflies entire surface

Payload Capabilities

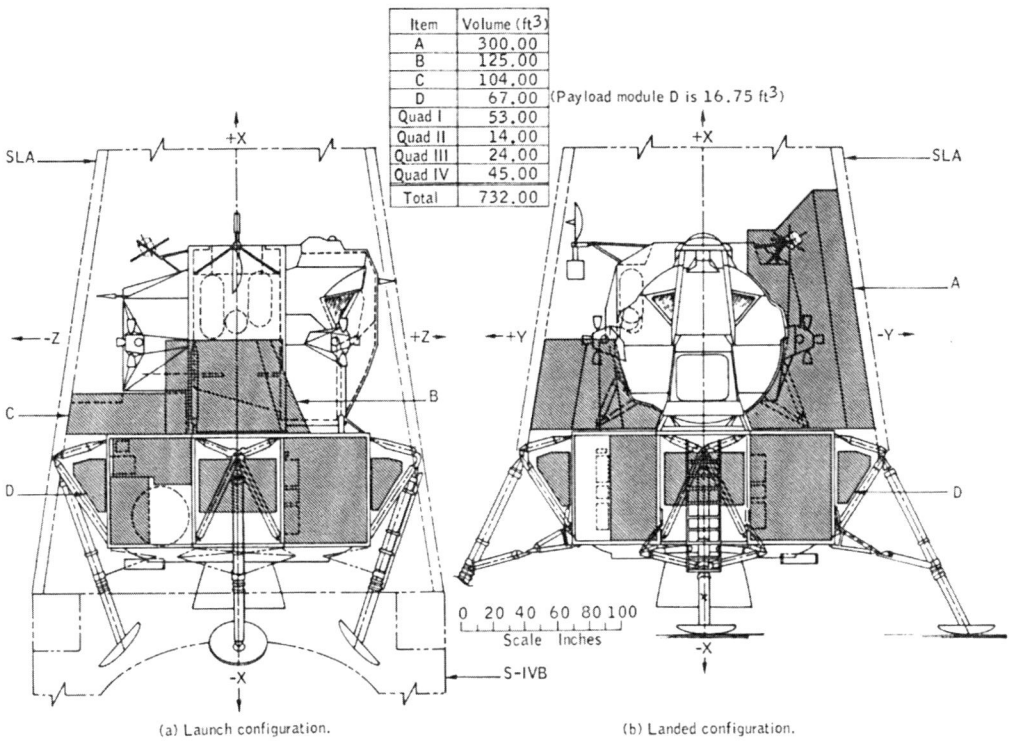

(a) Launch configuration. (b) Landed configuration.

Figure VII-1.- Usable volume of the lunar module.

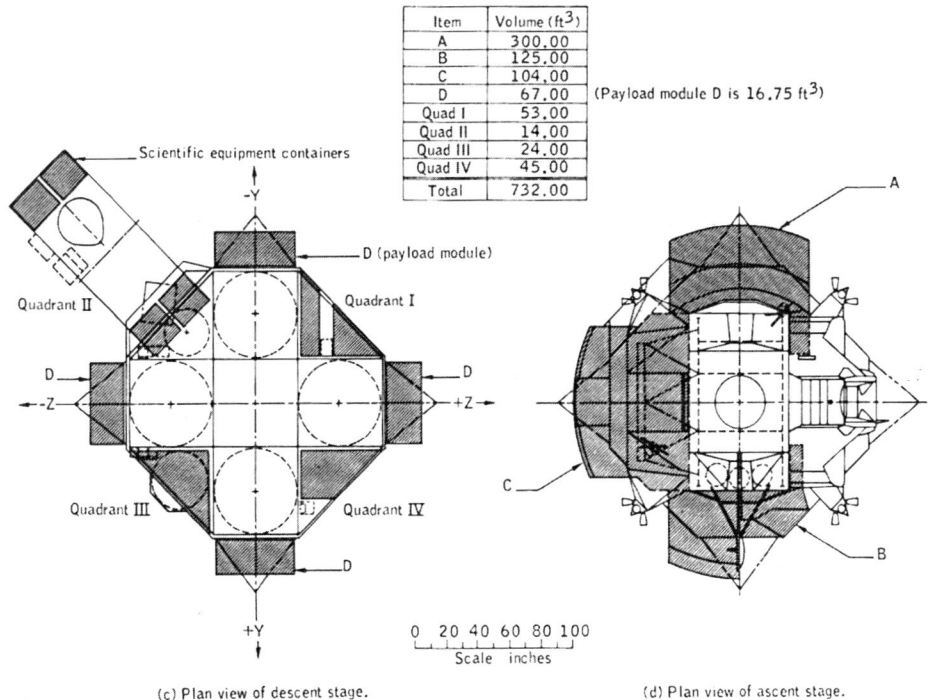

(c) Plan view of descent stage. (d) Plan view of ascent stage.

Figure VII-1.- Concluded.

SECTION VIII

OPERATIONAL CONSIDERATIONS

There are many factors that affect lunar mission planning. This section presents a brief description of some of the operational aspects and how they might influence lunar landing missions and payloads carried on these missions.

Section XV considers prelaunch and launch into earth orbit. One or two revolutions are required in earth orbit prior to translunar injection for equipment checkout, inertial measurement unit (IMU) alignment, and ground tracking for precise orbit determination. This period in earth orbit should not be overlooked.

Flight Times and Mission Duration

Translunar flight times from 60 to 132 hours are currently being considered. Flights of less than 60 hours appear unreasonable because of the very large propulsive requirements associated with them. The longer flight times appear desirable because of the lower propulsive requirements and because of the possibility of obtaining more favorable lunar orbit orientation (for some landing sites). The flight time variation allows mission flexibility in terms of initial orbit orientation (longitude of ascending mode) for high inclined lunar orbits and in terms of earth-launch window (primarily for :low inclined lunar orbits) when lunar landing lighting requirements are specified.

The total mission duration (earth launch to reentry) is limited by the spacecraft lifetime capability. For example, if a 14-day command and service module (CSM) of current design is used on a lunar polar mission that requires 7 days in orbit, only 168 hours is available for the total of translunar and transearth flight times. Obviously, the flight times allowable with this spacecraft would be more restricted for this example.

Multiple-Burn Lunar Orbit Insertion

Multiple service module (SM) engine burns were considered for lunar orbit insertion from the translunar flight. Some lunar orbit geometry configurations require as much as 58 000 pounds (9600 ft/sec Delta-V) of SM propellant for lunar orbit insertion when only a single SM engine burn is used. (Present SM propellant tanks hold only 38 000 pounds.) This value can be reduced to about 36 400 pounds (4900 ft/sec Delta-V) by two-burn techniques and to about 32 800 pounds (4300 ft/sec Delta-V) with three burns. The propulsive advantages of the multiple-burn technique are obvious; however, the technique does complicate the mission by requiring additional SM burns. The multiple-burn technique offers little or no propulsive advantage for low inclined lunar orbits.

The three-burn lunar orbit insertion is illustrated in figure VIII-1. The intermediate ellipse periods being considered vary from 6 to 30 hours, with corresponding apocynthion altitudes between 2500 and 10;250 nautical miles, respectively. The long-period ellipses will require navigation and possibly an additional midcourse correction on the downward leg. There should be ample time, however, to conduct experiments related to data that might vary with altitude.

The two-burn lunar orbit insertions fall into several classes, one of which is very similar to the single-burn insertion and one of which is similar to the three-burn insertion in that a large ellipse is involved.

It would seem, operationally, that the fewest possible burns would be desired. In other words, a mission would be flown with a single-burn lunar orbit insertion whenever possible. Extensions to two burns and three burns would follow when specific landing sites and conditions required increased propulsion.

Landing Flight Modes

Two flight modes are considered for landing. Mode I is the nominal Apollo profile beginning in an 80-nautical-mile-altitude circular lunar orbit. The lunar module (LM) separates from the CSM at 80 nautical miles and, by way of Hohmann transfer, descends to 50 000 feet where a powered descent is made to the landing site.

In mode 2, the CSM carries the LM down (from 80 nautical miles) into a low orbit such as a 50 000-foot circular orbit, where the LM separates and performs a powered descent to the landing site. Rendezvous occurs in approximately a 40-nautical-mile circular orbit. This mode is useful when additional lunar surface payload is desired and when additional SM propulsive capability is available. The LM descent tanks are full; therefore no payload above the nominal Apollo payload can be landed when mode I is employed. Mode 2 removes some of the propulsion requirements from the LM (allowing it to carry more payload) and increases the CSM propellant requirements by about 1400 pounds (200 ft/sec 0). Mode 2 allows an increase in surface payload of about 1200 pounds.

Landing Accuracies

Table VIII-I shows the landing accuracies that can be expected for several modes of operation.

The first six modes are based on preprogramed inertially controlled descent without; any landing point redesignation. These are the landing accuracies that are obtained without the astronaut in the loop and with only LM onboard navigation used. Mode 7 depends upon prior knowledge of the site and of LM computer reprograming of the surface altitude profile. Also, mode 7 accuracies are obtained without a crewman in the loop.

Improvements in landing accuracy are obtained in manned landings by use of the landing point redesignator. Landing accuracies similar to those obtained with small helicopters are achievable if sufficient LM propellant is available. Approximately 140 pounds (90 ft/sec Delta-V) of LM descent propellant is budgeted for landing point redesignation in, the nominal Apollo budget. If landing point redesignation is initiated at an altitude of 7000 feet, this propellant is sufficient to change the landing point a distance of about 1 nautical mile radially from the inertially guided point. The down-range and left-range designations are more easily obtained, and range shorting is precluded. Landing point redesignation initiation at lower than an altitude of 7000 feet results in less achievable change of the landing point. By proper targeting, suitable Delta-V budgeting, and landing point redesignation, a manned vehicle should have a high probability of landing within 100 feet of a previously landed vehicle.

Landing Site Latitude and Longitude Effects

The above discussion was concerned with the accuracy of landing the LM at a selected site, assuming the site was accessible to the spacecraft. The effects of landing site latitude and longitude can be seen by studying figure VIII-2, a typical example depicting the lunar landing sites which are accessible within the 14-day CSM capability. Both site latitude and longitude must be considered when investigating landing capability. For the I- and 4-day surface staytimes, all latitudes, including the poles, are accessible if the appropriate longitude is chosen. Figure VIII-2 is based on a standard Apollo-type landing and payload capability. Multiple SM burns are used for lunar orbit insertion and transearth injection. The contours in the figure are spacecraft-capability limited (abort included) and, as such, the accessible area must decrease with increasing payload. It should be noted that the 7-day surface staytime is very limited, primarily by abort and rendezvous considerations. These accessible areas are shown in figure VIII-3 as they appear on the lunar disk when viewed from the earth.

Surface Staytime

Figures VIII-3 and VIII-4 also indicate the effects of surface staytime on accessible area. The longer surface staytimes are restricted primarily by the surface-to-orbit abort and rendezvous requirements, while the short staytimes are limited primarily by the spacecraft capability to insert into and inject from high inclined lunar orbits.

It should be emphasized that a 14-day lunar surface staytime requires both an extended CSM lifetime and landing sites very near the lunar equator. Figure VIII-4 indicates the area accessible for a 14-day surface staytime when using a 23-day CSM.

Abort Considerations

The surface staytime, landing site latitude, and abort considerations are closely related through the geometric relation between the CSM orbit and the landing site. Figure VIII-5 illustrates a CSM orbit track and the relative motion (due to the moon's rotation) of the landed vehicle. Increasing the surface staytime requires increasing the CSM orbit inclination; also, increasing the landing site latitude increases the required CSM orbit inclination. CSM orbit inclination, along with several other factors, is of prime consideration to the propulsion requirements for lunar orbit insertion and transearth injection. If abort capability from lunar orbit to earth is required for each orbit revolution, the CSM orbit inclination necessary for the desired surface staytime can severely restrict mission capability.

Abort from, the lunar surface to the CSM is also affected by the orbit inclination. The out-of-plane propulsion requirements at rendezvous are determined by the distance between the LM at launch and the CSM orbit track as shown in figure VIII-5. This means that the maximum orbit inclination will be restricted by the landing site latitude and by the propulsion capability at rendezvous.

In general, increasing the surface staytime increases the abort propulsion requirements for both surface-to-orbit and lunar orbit-to-earth aborts. Surface staytimes of greater than about 7 days are achievable only for near equatorial landings, if the abort-each-revolution criterion is maintained. Long surface staytimes require either near-equatorial landings or relief of the any-orbit abort requirement. Relieving the abort requirement usually results in excessively long periods in which no abort is possible.

A distinction should be made between any-orbit abort and any-time abort. Any-orbit abort implies that abort can be done at least one time during any orbit, whereas any-time abort infers abort at any instant during any orbit. Any-time abort does introduce either a propulsion penalty or an increase in orbit waiting; time for proper relative positioning of the two vehicles. This phasing time can be as long as 18 hours; therefore, it might be desirable to wait, on the surface for the next CSM revolution (2-hour maximum). It is possible to conceive of situations in which it would be best to abort immediately and then wait long periods in orbit before rendezvous. For example, a sizable leak in the ascent propellant tanks or a fire in the descent stage would be immediate-abort situations. Some missions would be surface-payload penalized if the any-time-abort requirement were maintained. Careful consideration should be given to the value of the additional payload (the amount of which is dependent upon the specific mission) as compared to the desirability of any-time abort.

The preceding paragraph is associated with aborts from the surface-to-lunar orbit only. Aborts from lunar orbit to earth are practical only once per revolution.

Lunar Surface Lighting

Lunar surface lighting conditions are of concern in three areas: the landing maneuver, extravehicular activities (EVA) on the surface, and the spacecraft thermal environment.

Simulations have been conducted. for lunar-landings under a wide range of lighting conditions, including those to be expected from earthshine (ref. 1). The results indicated that landings were practical under lighting equivalent to three-fourths earthshine. Earthshine landings introduce the possibility of increasing the earth-launch opportunities. The desirable earth elevation (similar to sun elevation) at landing is 70 to 150 above the lunar horizon and behind the astronaut. If this earth. elevation requirement is maintained, the possible earthshine landing sites will be severely restricted by the earth-moon geometry.

Some question exists about astronaut EVA capability under earthshine lighting conditions. While earthshine is many times brighter than full moonlight at earth, recent tests indicate that an astronaut may have difficulty in walking on the unfamiliar lunar surface. There is also some question of astronaut EVA capability under full-sunshine conditions. This document makes no effort to determine which lighting condition is more desirable.

The earthshine thermal environment imposed on the LM is very similar to total darkness (-250° F) since practically no heat is received from the earth. Potential problems include freezing of the LM ascent propulsion system, which is currently qualification tested to only 400 F. The addition of heaters and batteries would result in a heavier LM ascent stage.

Mobility Aids

Two basic mobility aids are being considered: lunar surface roving vehicles and lunar flying vehicles. These vehicles are described in section X. Both vehicles require astronaut activity in deploying, preparing, and operating the units. These activities place additional time and effort requirements on the crew.

The roving vehicles offer close, detailed inspection of the path traversed and require considerable time to make the traverse (average speed is about 5 miles per hour). Therefore, the roving vehicles require more surface staytime than the flying vehicles.

The operating range of the roving vehicle will be less sensitive to the payload weight than the flying unit, since the range of the flying unit will be sensitive to the weight it must carry.

The flying vehicles offer the advantage of rapid traverse, if investigation of particular points is desired. Some of the flying units under consideration depend on the use of LM descent propellant residuals. The amount of residuals available will depend on piloting ability and required landing accuracies and will, therefore, not be known until the vehicle is on the lunar surface. This delay in information could present problems in the effective planning for the use of the flying vehicle.

Both the flying and roving vehicles will probably be range restricted by crew safety considerations. If either vehicle fails, they crewman will either have to walk back to the LM or be rescued by his crewmate. Current thinking involves either carrying a flying vehicle for rescue or restricting the mobility vehicle range to a safe-walk-back distance, as indicated in section X.

Communications

Communications can impose operational constraints. the operational range of the mobility vehicles will be restrained by the necessary line of sight between the LM and mobility vehicle antennas. This range is about 4 miles (LM antenna, 40 feet high) on a smooth, spherical moon. Any obstructions, such as a crater wall or a small hill, will greatly reduce this range:

The LM to CSM communications appear to present, no problems other than those problems which might exist if a landing is made very close to an obstruction. Crater landing sites may reduce the amount of communication time available during each revolution. About 15 minutes of communication time is available each revolution for the nominal Apollo mission. The 80-nautical-mile orbit requires 2 hours to make a revolution.

The CSM to earth communications are dependent on the CSM lunar orbit orientation. Communication times between 68 and 120 minutes (100 percent of the time) are obtainable each revolution, depending on the CSM lunar orbit orientation.

The LM to earth communications exist any time a line of sight to earth exists. This indicates that landings on the lunar backside will require some form of communications relay. In general, the lunar near-earth side will have LM to earth contact with possible problems near large obstacles. Some potential landing sites require LM orientations during the powered descent that do not provide communications throughout descent.

TABLE VIII-I.- LUNAR MODULE LANDING DISPERSION SUMMARY

Mode	Changes required	LM configuration	LM powered descent initiation altitude, 1σ values	
			80 n. mi.	50 000 ft
1	None	Manned LM	σ_R = 3100 ft σ_L = 2400 ft	σ_R = 2680 σ_L = 2400
2		Unmanned LM	σ_R = 3100 ft σ_L = 2400 ft	σ_R = 2680 σ_L = 2400
3	Software only	Manned LM, reticle fix before powered descent	Not effective	σ_R = 900[a] σ_L = 1650
4		Unmanned LM, reticle fix before powered descent	Not effective	Not enough time
5		Manned LM, fix from CSM over landing site	Not possible	σ_R = 900[a] σ_L = 1650
6		Unmanned LM, fix from CSM over landing site	Not possible	σ_R = 900[a] σ_L = 1650
7	Software and hardware	Range-to-go steering a priori range information in altitude profile	σ_R - 500 ft[b] σ_L = 2400 ft	σ_R - 500[b] σ_L = 2400
8		LM homing on beacon on LM truck	<100 ft[a]	<100[a]
9		Television command system remote LPD from earth	<1000 ft[a]	<1000[a]

[a] Estimate.
[b] Very preliminary number — actual answer dependent on terrain model, et cetera.

Operational Considerations

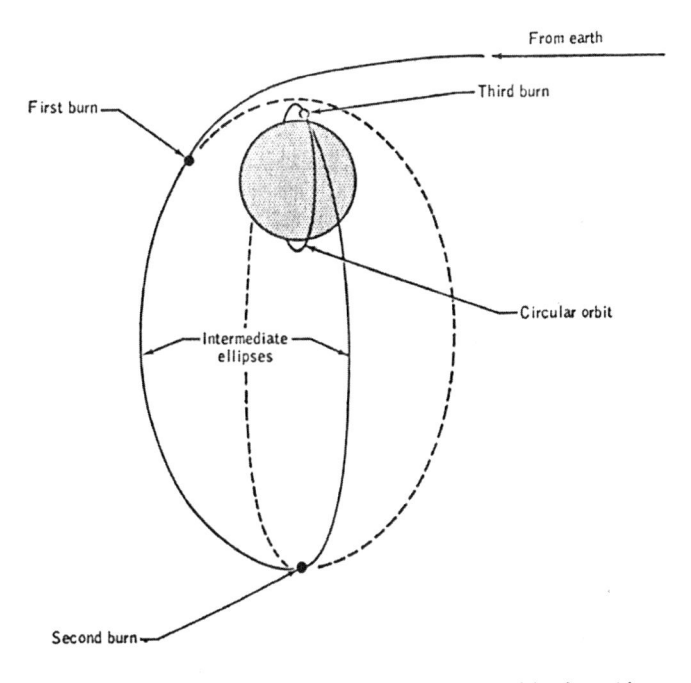

Figure VIII-1.- Three-impulse geometry for lunar orbit insertion.

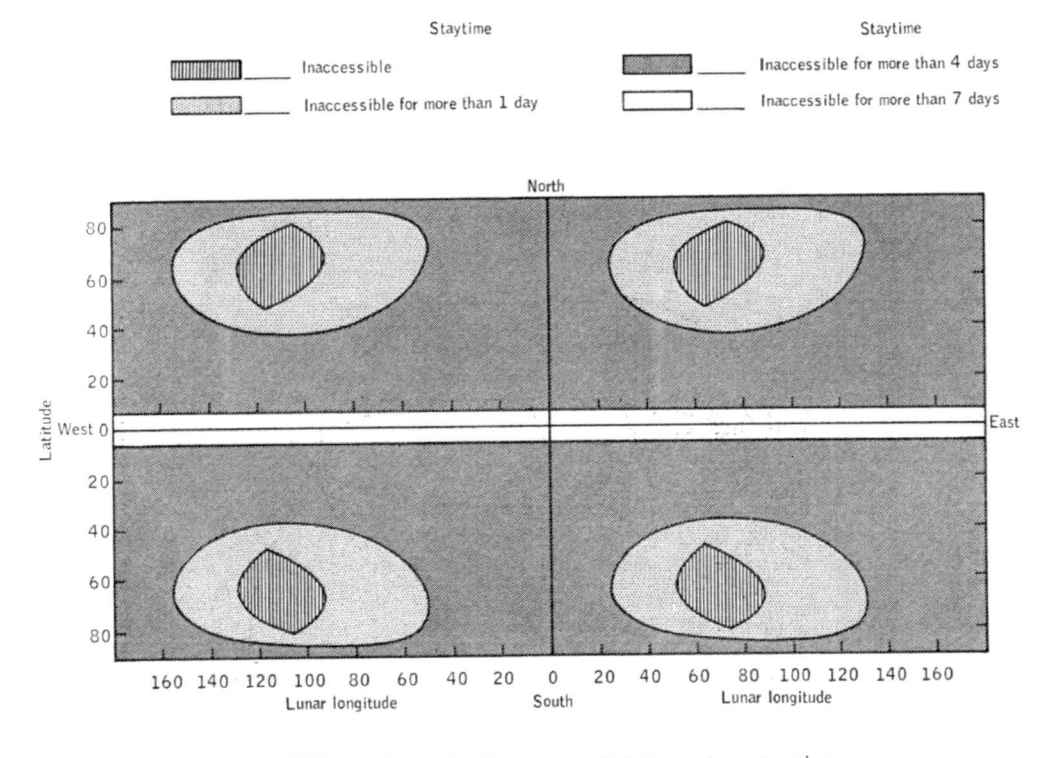

Figure VIII-2.- Lunar landing accessibility using the 14-day command and service module.

Figure VIII-3.- Lunar landing accessibility using the 14-day command and service module, as viewed from earth.

Figure VIII-4.- Lunar landing accessibility using the 23-day command and service module, as viewed from earth.

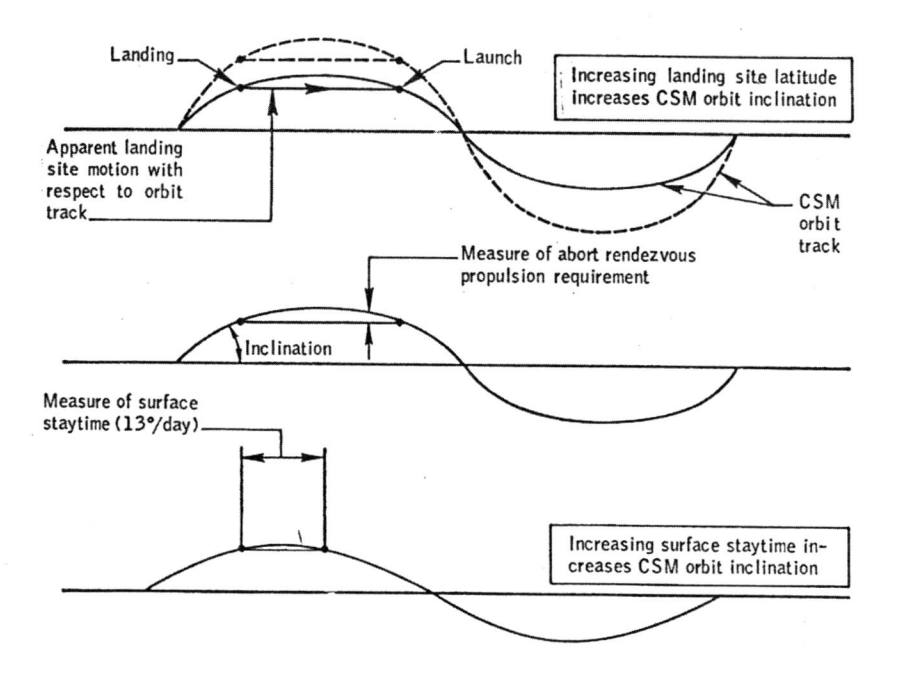

Figure VIII-5.— Lunar orbit geometry for staytime, site latitude, and abort considerations.

SECTION IX

STAYTIME EXPENDABLES

Staytime expendables are of interest because they reduce the amount of payload available for landing scientific equipment and for providing mobility on the lunar surface. For example, the addition of expendables and associated fixed hardware to extend the capability of an augmented lunar module (ALM) from 2 to 14 days would reduce the capability to carry scientific payload by approximately 2050 pounds. Figure IX-1 can be used to approximate increases in expendable mass as staytime is increased above the 2-day staytime provided by the standard LM.

Figure IX-1, which can be used for the extended lunar module (ELM) or for the ALM, is based on the addition of a solar array; of environmental control system (ECS) radiators; and of required tankage and consumables such as water, oxygen, lithium hydroxide, and food at the required rates. Backup information for figure IX-I includes the following.

Expendables Versus Days Staytime

	Item	Approximate weight, lb
1	Basic LM expendables	300
2	Solar array (1 kW)	200
3	ECS radiator	300
4	Cryogenic O_2 tank — first unit	50
5	Cryogenic O_2 tank — additional units	100
6	H_2O tank	50
7	Daily expendables (four depressurizations)	100
	Oxygen	40
	Water	30
	Food	5
	Lithium hydroxide	25

To determine the expendable plus the fixed hardware mass increase over the standard LM, the following data are used.

Days staytime	Items[a]	Mass increase over LM, lb
3	2 + 3 + 4	550
4	7	650
5	7 + 5	850
6	7	950
7	6 + 7	1100
8	7	1200
9	5 + 7	1400
10	7	1500
11	7	1600
12	7	1700
13	5 + 6 + 7	1950
14	7	2050

[a] Item numbers correspond to the item numbers in the first table and are additive as staytimes increase.

Life Support Requirements

The consumables required for use on the lunar module shelter (LMS), early lunar shelter (ELS), staytime extension module (STEM), and lunar module truck shelter (LMTS) may be up to 45 percent less than required on the ELM or ALM because of differences in vehicle leak rates, use of airlocks, and method of power generation (fuel cells or solar arrays).

Figure IX-2 provides a comparison of LMS oxygen usage with and without an airlock, as well as with various modes of airlock operation. It can be determined from figure IX-2 that incorporation of an airlock is not desirable on the LMS from a mass standpoint unless 14 or more extravehicular sorties are planned for a given mission and unless it is anticipated that the cabin will be pressurized during extravehicular activity (EVA). The airlock configuration considered in the past will only accommodate one man; thus, if lunar operations required two men to leave the ALM at the same time, the number of airlock decompressions with an airlock would be three times the number of decompressions required without an airlock, assuming an unpressurized cabin during EVA.

Configurations, such as STEM, ELS, and LMTS provide airlocks with two-man capability. The larger vehicle volumes of these configurations would be double the number of extravehicular sorties.

Electrical Power Generation

Power generation for lunar shelters can be accomplished by several different means, depending on the specific mission involved. In general, any given shelter could be powered by solar array or fuel cell systems. Batteries, as a prime source of power, are not generally economical from a mass standpoint for periods in excess of 2 or 3 days. Power generation systems using solar arrays have a mass of approximately 200 pounds per kilowatt of generating capacity. The reliability of solar arrays is considered to be high, and the arrays can be relatively easily sized to match power generation requirements. Solar arrays are limited to lunar day operations.

Fuel cell power systems are suited to lunar day, lunar night, or earthshine operations; however, fuel cell systems must be redundant to some degree to provide the desired reliability and are not as readily sized to mission requirements. A steady-state 2-kilowatt fuel cell system would have a fixed mass of approximately 170 pounds and would require approximately 3.0 pounds of reactant and tankage per kilowatt-hour of power generated. For every kilowatt-hour of power generated, 0.80 pound of water would be produced.

Power generation systems, as well as consumable tankage and storage space, are in general provided on a given shelter; however, in certain cases a modular concept may be required or be more desirable where these items are packaged and carried to the lunar surface on an unmanned logistics-type vehicle and probably transported to the shelter on a surface vehicle of the local scientific survey module (LSSM) type. An example of this system is shown in figure IX-3

Command module (CM) consumables can be provided in the equipment bay of the service module (SM) for missions of up to 10 days, which would allow a lunar staytime of approximately 4 days. Where lunar staytimes in excess of 4 days are required or where trip times are increased, consumables would have to be added at a rate of approximately 85 pounds per day, plus tankage. A portion of the bay reserved for scientific equipment would be used as storage for the extra consumable items.

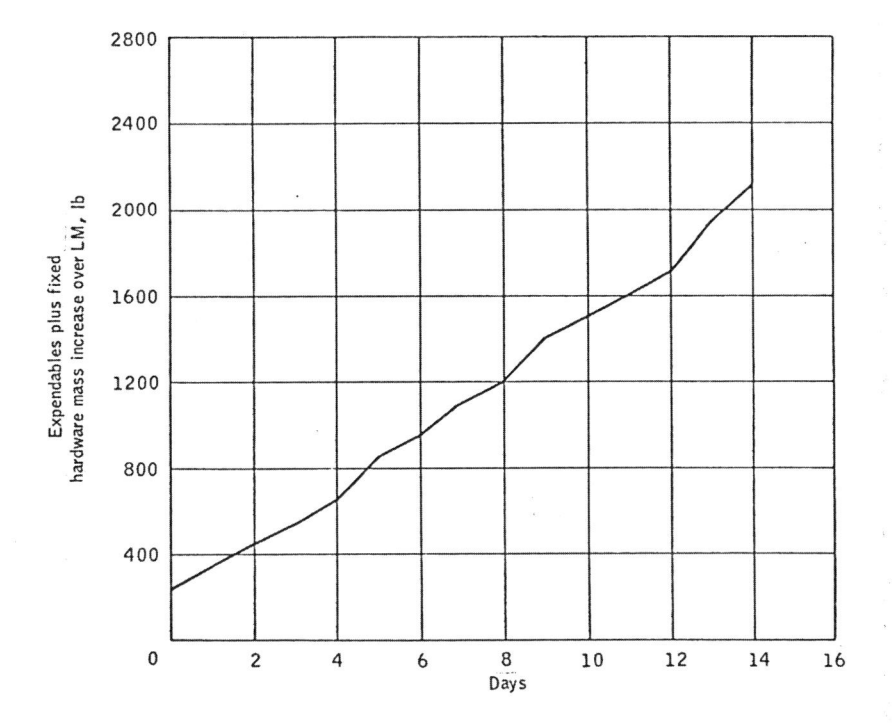

Figure IX-1.- Expendables versus days staytime.

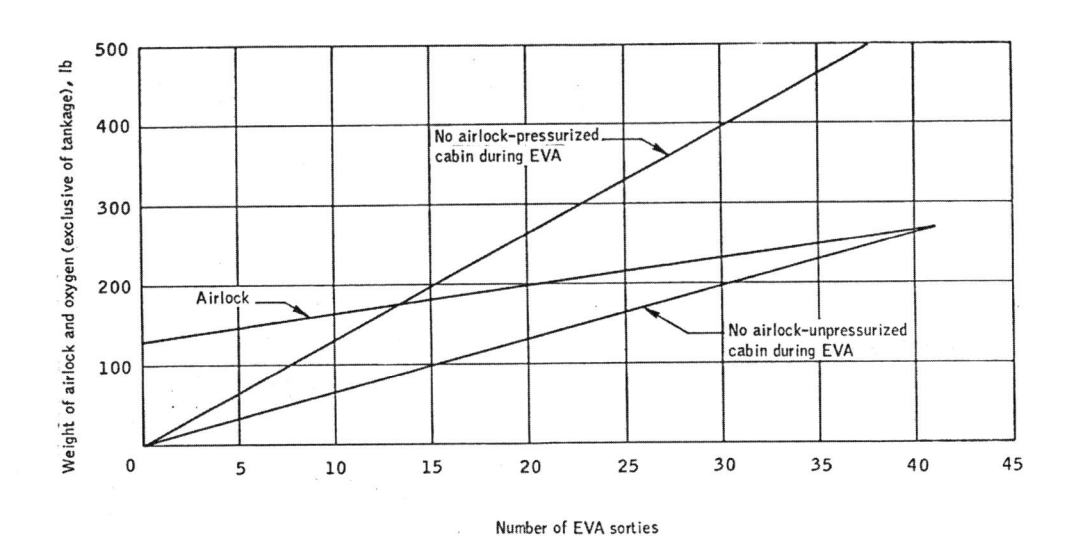

Figure IX-2.- Weight trade-off airlock/no airlock versus number of EVA sorties.

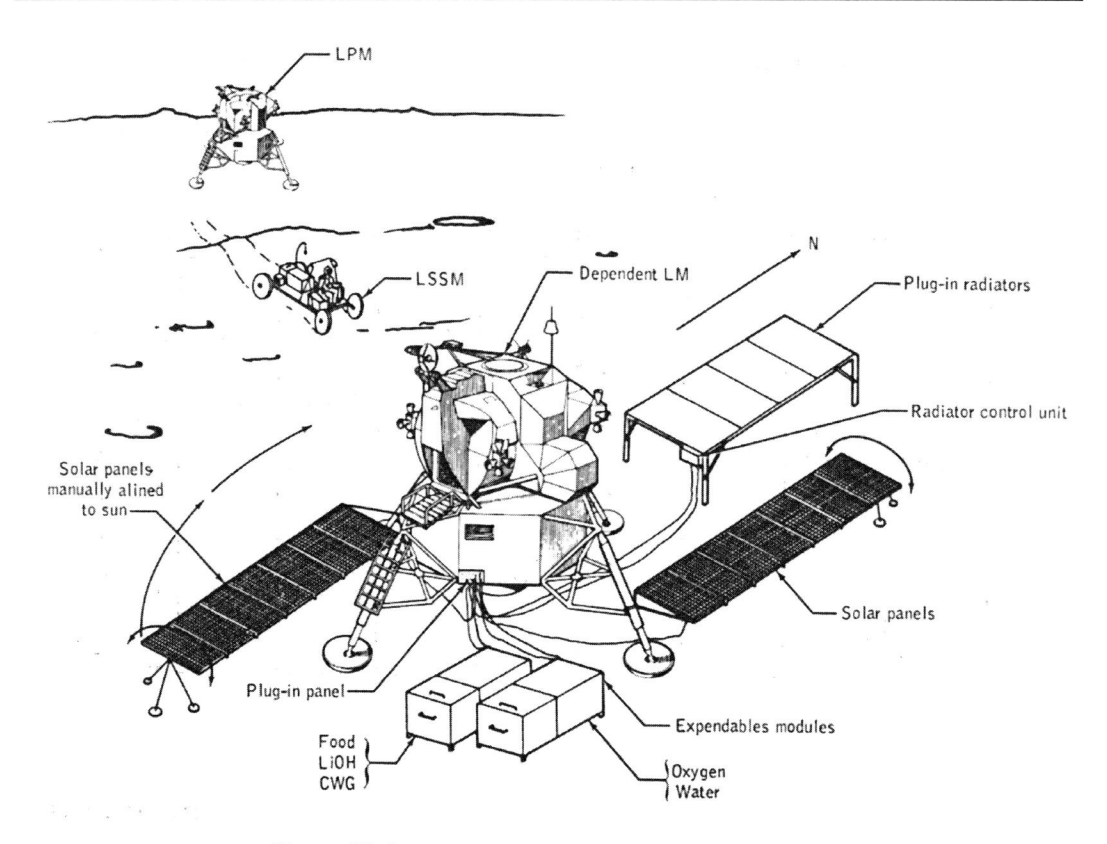

Figure IX-3.- Modular mission extension provisions.

SECTION X

LUNAR MOBILITY

Lunar mobility aids are used to enhance the lunar surface exploration capability. Two basic methods, rovers (or wheeled vehicles) and flyers, are under consideration.

Lunar Roving Vehicles (Manned)

Several types of manned roving vehicles for lunar surface mobility have been envisioned and studied. These vehicles have ranged in size and utility from the local scientific survey module (LSSM) which is a short-range, dependent, open-type vehicle to the mobile exploration (MOBEX), mobile laboratory (MOLAB) type vehicles which are long range and self-sustaining and have environmentally controlled crew areas.

Local Scientific Survey Module

The LSSM is a battery operated, wheeled, surface-mobility aid for short-range transportation of men and equipment on the lunar surface. The vehicle has a gross operating mass of approximately 2000 pounds which includes a suited astronaut with his associated life support equipment and 700 pounds of scientific equipment. The basic vehicle weighs approximately

1000 pounds. The vehicle has a total range of 15 miles per sortie with a total mission traverse distance of approximately 125 miles. The batteries used to power the vehicle are rechargeable from a separate source and can be recharged within an 8-hour period after each typical 15-mile sortie. The LSSM, because of its mass, is dependent upon a dual launch mission with the delivery being made by a lunar payload module (LPM), a lunar module shelter (LMS), or an LM truck-type vehicle. The two LSSM designs under study are shown in figures X-1 and X-2. Stowage for delivery is shown in figure X-3. The performance requirements are shown in the following table:

```
Scientific equipment capacity  . . . . . . . . . . . . . . . .  700 lb

Operational period . . . . . . . . . . . . . . . . . .  Lunar day only

Total traversed distance . . . . . . . . . . . . . . . . . .  125 miles

Maximum sortie range . . . . . . . . . . . . . . . . . . . .  15 miles

Minimum speed (soft soil, level) . . . . . . . . . . . . . . .  2.5 mph

Obstacle negotiability (very low speed) . . . . . . . . . . .  12 in.

Crevice negotiability  . . . . . . . . . . . . . . . . .  20-in. gap

Slope negotiability  . . . . . . . . . . . . . . . . . . . . . .  35°
```

The LSSM, as presently specified, does not have a communications or navigation system. The astronaut is furnished his communications and life support systems through the portable life support system (PLSS). Navigation is accomplished by lunar maps and by an odometer installed in the vehicle.

The first flight article can be ready for delivery to NASA approximately 3 years after go-ahead. This first vehicle would be ready for launch approximately 6 to 8 months later. The earliest a vehicle could be used on a lunar surface mission is mid-1971. This coincides with the availability of the LPM or LM truck-type delivery vehicles. The nonrecurring costs are approximately $40 million with recurring costs of $4 million.

Mobile Exploration, Mobile Laboratory Vehicles

The MOBEX, MOLAB vehicles are vehicles electrically powered by fuel cells or nuclear radioisotope thermoelectric generator (RTG) power systems having mission times from 14 to 90 days. These vehicles are designed primarily for long traverses from 250 to 2000 miles depending on the vehicle and the staytime. They are essentially mobile bases with laboratories and a living space which is environmentally controlled and are self-sustaining for the full length of the mission.

The gross operating mass of these vehicles ranges from 7000 to 18 500 pounds and would depend on an LM truck-type lander for vehicle weights up to 12 000 pounds and on a new type direct lander for vehicles of a larger size.

The availability of the MOBEX, MOLAB vehicle would be the 1974 to 1975 time period with a lead-time of 5 years required to the first mission article. The development cost (nonrecurring) would be approximately $350 million with a recurring vehicle cost between $15 million and $18 million including launch support costs.

Lunar Flying Machines

A review of the overall lunar program, of the study of payloads, and of the capabilities required on various missions indicates a need for a lunar flying vehicle on early lunar landings. Lunar flying vehicles weighing approximately 150 to 440 pounds can be delivered to the lunar surface on single-launch missions, as well as on dual-launch missions. This vehicle will significantly increase the capability to investigate a number of scientifically interesting sites and to serve as a rescue vehicle. Other general capabilities include reconnaissance with or without a camera, exploration of rough terrain and craters, and traverse over inaccessible terrain requiring a minimum of time.

The propellant for the flying vehicles would be the LM descent-stage residuals on the single launches, the LM descent-stage residuals of logistic deliveries on dual launches, and additional propellants carried as part of the lander payload. Assuming no LM system malfunctions, but including engine off-nominal operation and worst-case intertank head differential and zero hover time, the LM residual propellant is estimated at 1416 pounds. With optimum engine and system performance, zero head differential, and no hover time, the residual propellant is estimated at 1603 pounds. The maximum hover or descent time from 500 feet above the lunar surface to the surface is 110 seconds, and the nominal hover time is estimated at 55 seconds. The LM uses 9.3 lb/sec of propellant so that the maximum LM residuals using nominal hover would be 1091 pounds, and using the 110 seconds of hover plus the nominal case of LM residuals of 1416 pounds a minimum of 393 pounds of propellant would remain.

Table X-I lists flying vehicles investigated with the ranges and propellant requirements for one astronaut plus 100 pounds of payload and one intermediate stop at the exploration site.

Figure X-4 is representative of vehicles 1, 2, and 3 and figures X-5 and X-6, vehicles 4 and 5 in table X-I. Several curves follow and the numbers on the curves correspond to the vehicle number tabulated in table X-I per load of ;propellant. Figure X-7 shows flight time for the vehicles versus radius or one-half the total distance in miles. Figure X-8 is a plot of the total flight distance in miles versus flyer total weight, including the amount of propellants, the suited pilot, and a 100-pound payload for one intermediate stop from home base. Figure X-9 is a curve of optimum velocity versus flight distance in miles for a pilot and 100 pounds of payload. Deviation from the optimum speed can cause a decrease in overall flight distances. As an example, vehicle 2 has a range of speed from 320 to 500 ft/sec with the optimum at about 420 ft/sec. Operation at 250 ft/sec will reduce the overall range or distance by approximately 10 percent.

The larger weight vehicles listed in table X-I can be carried only on dual-launch missions, whereas the smaller weight vehicles can be carried on either single- or dual-launch missions.

As an example, vehicle 1 from table X-I is a simple, lightweight one-man vehicle having an approximate total flight distance of 15 miles with one intermediate stop and 100 pounds of payload for each load of propellants. Table X-II illustrates the total flight distance for other payloads and numbers of intermediate stops. The 300-pound payload could be scientific instruments, additional propellants for increased flying range, or an inactive, suited astronaut. Since this vehicle carries a 230-pound load of propellants, an additional load of propellants can be carried, increasing this flyer range. From table X-II, a 300-pound payload (230 pounds of propellants) will provide a flying radius of approximately 15.7 miles, and by switching to a set of tanks or propellants carried as a payload, will allow a return trip of 15.7 miles with another 300-pound payload.

Using the LM residual propellants from one descent stage, a minimum of three and a maximum of six trips or sorties could be accomplished, depending on the amount of residuals. It is possible to transport, two of these flyers to the lunar surface on the manned LM vehicle and always retain one at the LM for rescue operations.

Lunar surface ultrahigh frequency (UHF) communications are presently limited to approximately 4 miles. One way of extending communications is to use the normal PLSS backpack and to install a relay package or reflector approximately every 4 miles. For long traverses on a large flying machine it may be necessary to install a communication system on the flying vehicle for communications to earth either directly or by relaying to several satellites in orbit around the moon.

Navigation equipment will depend on the vehicle complexity, the traverse distance, and the detail mapping acquired prior to the flights carrying these vehicles.

TABLE X-I.- FLYING VEHICLES INVESTIGATED

Vehicle	Dry weight + residuals, lb	Usable fuel, lb	Maximum radius, miles
1	155	230	7.3
2	215	332	10.8
3	265	476	16.1
4	440	635	24.1
5	756	1480	50.0

TABLE X-II.- ONE-MAN FLYING MACHINE

[155-lb dry weight and 230-lb capacity]

Average flight payload, lb	Number of intermediate stops	Flight, km	Distance, statute miles
0	0	57.9	36.0
	1	33.8	21.0
	2	25.8	16.0
	3	21.4	13.3
	4	18.2	11.3
	5	15.5	9.6
50	0	40	24.9
	1	28.3	17.6
	2	21.5	13.4
	3	18.7	11.6
	4	15	9.3
	5	12.4	7.7
150	0	35	21.8
	1	21	13.1
	2	16	10.0
	3	13.2	8.2
	4	10.8	6.7
	5	8.7	5.4
300	0	25.3	15.7
	1	15.2	9.4
	2	12.2	7.6
	3	10.6	6.6
	4	8.5	5.3
	5	6.6	4.1

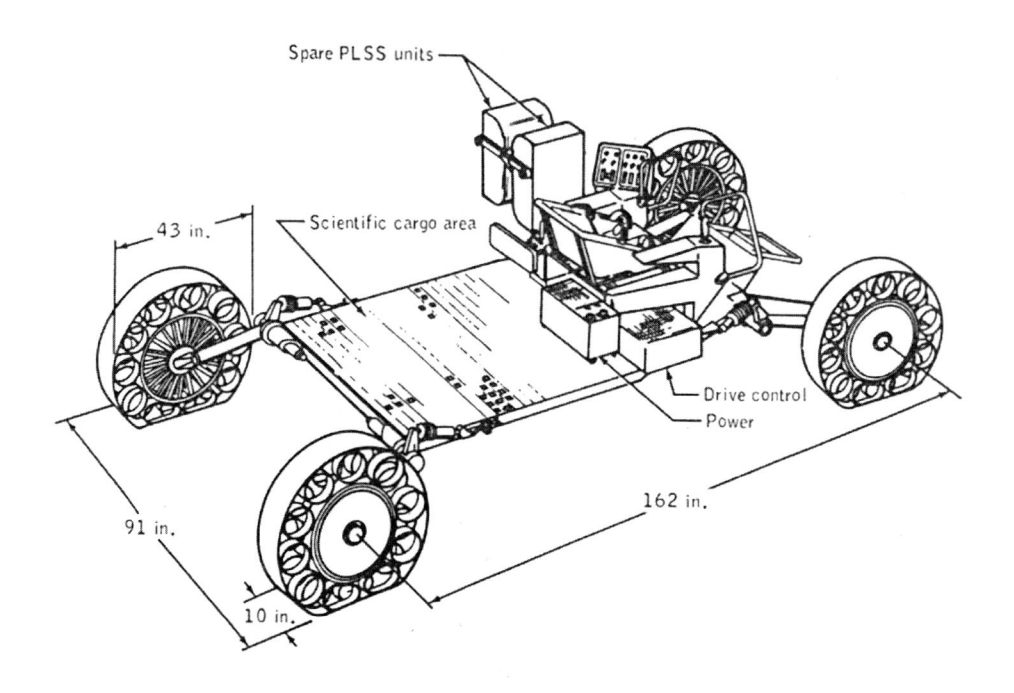

Figure X-1.- Local scientific survey module concept, four wheel.

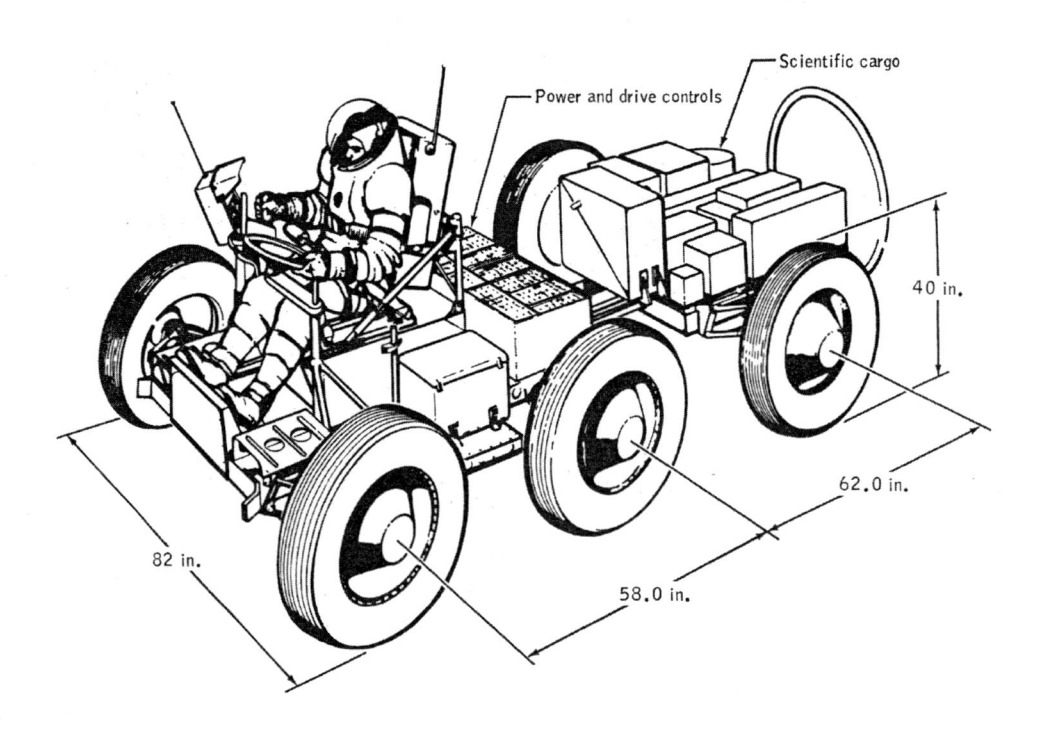

Figure X-2.- Local scientific survey module concept, six wheel.

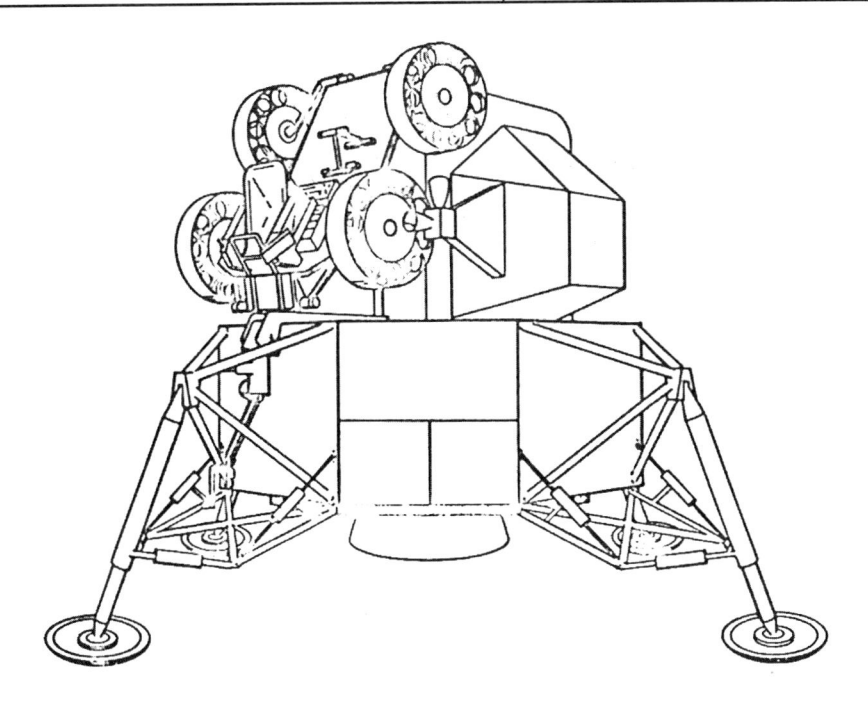

Figure X-3.- Lunar module shelter or stripped lunar module, lunar scientific survey module stowed configuration.

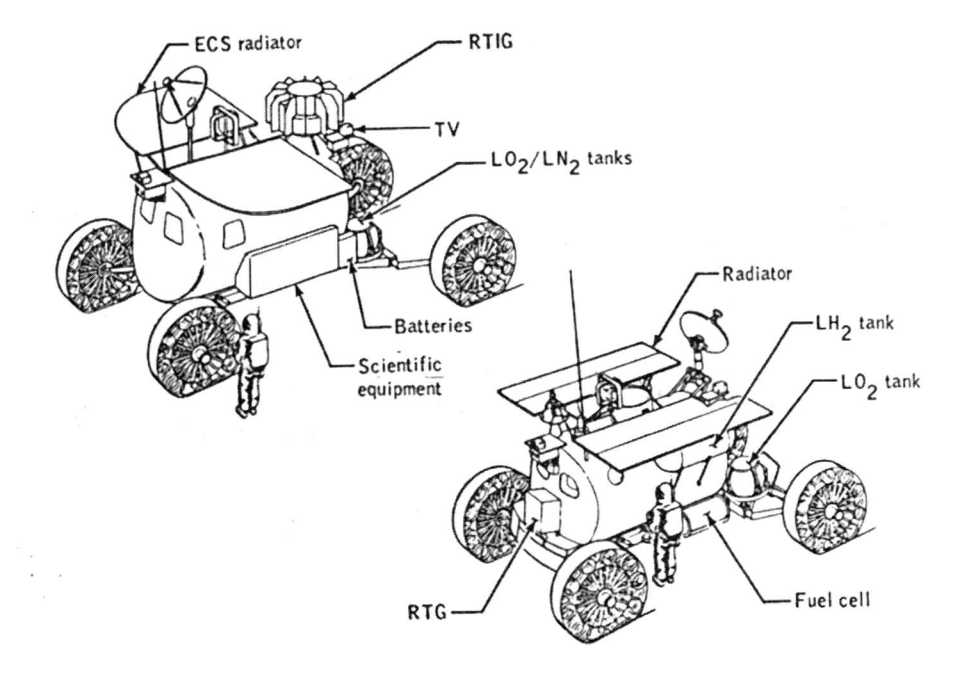

Figure X-4.- Mobile exploration, mobile laboratory type vehicles.

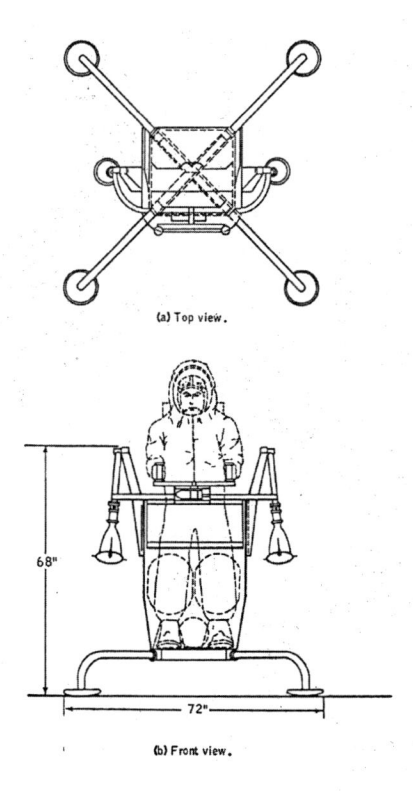

Figure X-5.- Lunar flying unit.

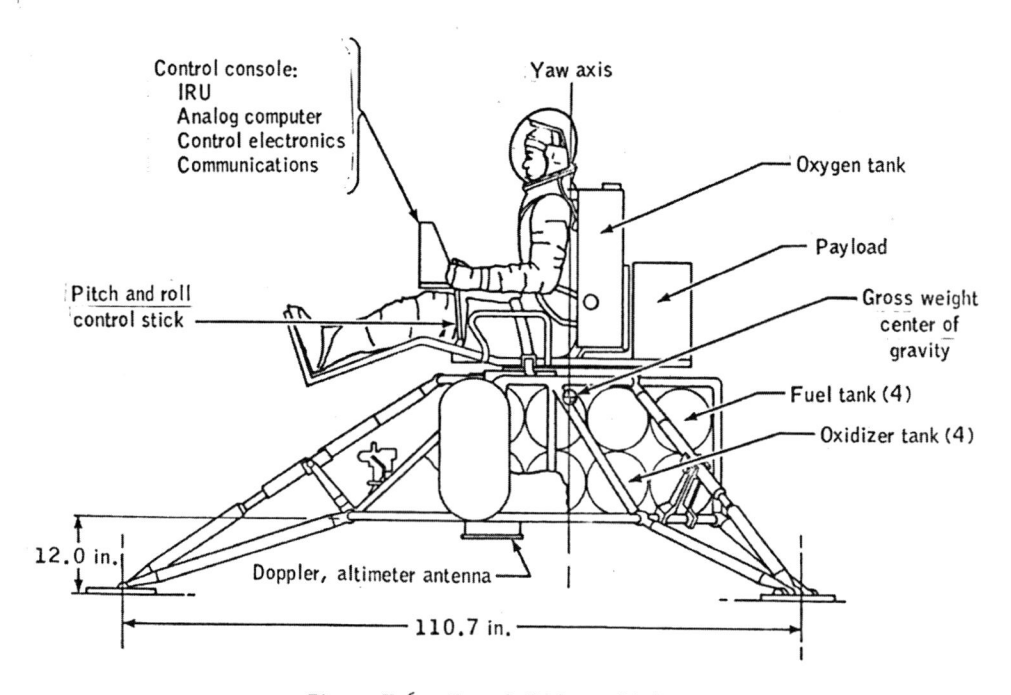

Figure X-6.- Manned flying vehicle.

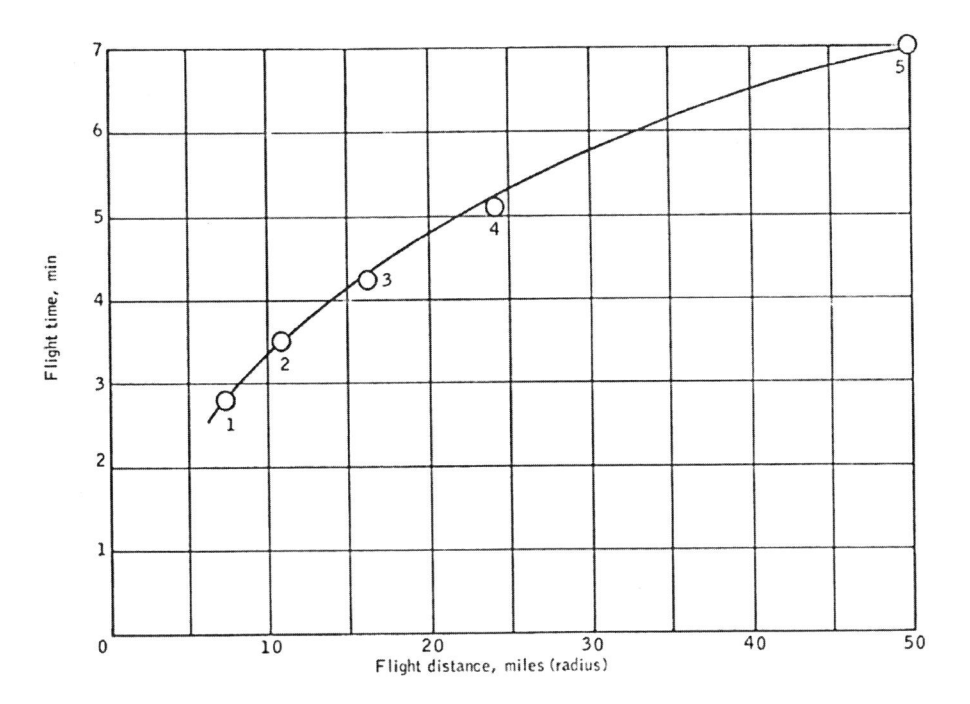

Figure X-7.- Flight time versus distance.

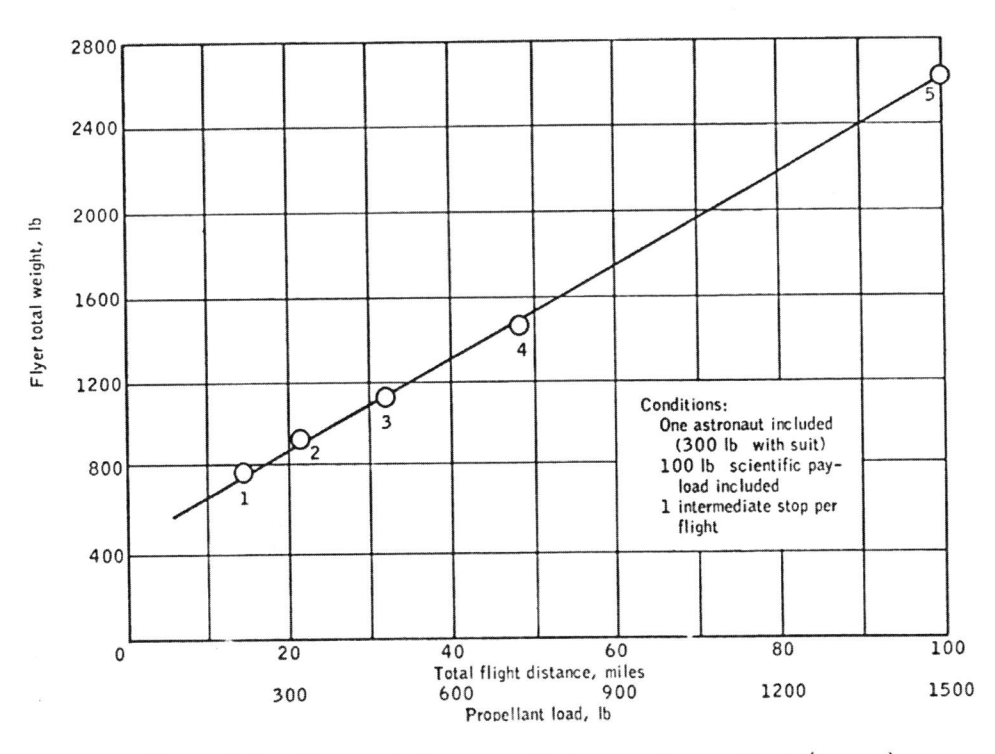

Figure X-8.- Flyer weight versus distance and propellant (manned).

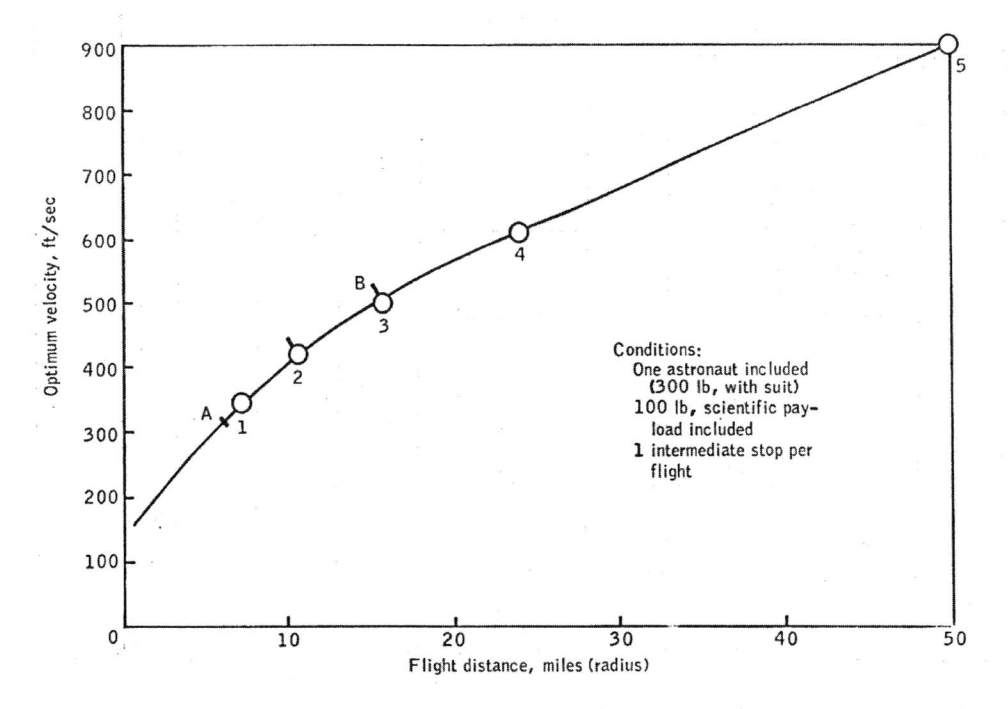

Figure X-9.- Flyer velocity versus distance.

SECTION XI

LUNAR EXTRAVEHICULAR ACTIVITY OPERATIONS

Certain items of protective and support equipment must be provided for support of an extravehicular activity (EVA) astronaut during lunar surface excursions and for performance of operational and experimental tasks. The basic suit hardware requirements for support of the physiological needs include the following:

1. Environmental protection (pressure integrity, meteoroid penetration, and thermal comfort)
2. Mobility and articulation (unencumbered)
3. Visibility (broad, unobstructed visual. field, plus eye protection)
4. Structural integrity and durability (wear, slash, tear, and abrasion resistance)

The basic portable life support system (PLSS) hardware requirements for support of the physiological needs include the following:

1. Breathing oxygen supply.
2. Carbon dioxide and thermal control
3. Pressurization control
4. Humidity and noxious gas control
5. Communications
6. Circulation
7. Emergency capability
8. Failure warning system

In addition to these basic equipment requirements, the EVA astronaut may have to wear or carry other equipment or specialized aids required by the mission. Initial lunar landing capabilities are developed around the following design requirements:

1. Twenty-four-hour lunar surface excursion capability (12 hours per crewman)
2. Four-hour capability for single excursions (3 + 1 hours contingency)
3. Range radius of up to I nautical mile
4. Day or night excursion protection (±250° F)
5. Total metabolic capacity of 4800 Btu (1200 to 1600 Btu/hr, average rates)
6. Capability for free space transfer from one vehicle to another and depressurized cabin operations

Space Suits

Apollo soft suit.- The Apollo space suit assembly consists of (1) a liquid coolant garment (LCG) which is much like a set of long underwear (fig. XI-1) that has a network of small flexible tubes attached (These liquid transport tubes cover a major portion of the body surface area (fig. XI-1) and provide the primary means of body cooling. Water from a support pack is pumped through the tubular network to carry body heat to the support pack heat exchanger, where it is rejected to space); (2) A pressure-retaining layer of polymer-coated gas-impermeable fabric; (3) A restraint layer to control the shape and configuration of the pressure-retaining layer and to contain the mobility system. The mobility system consists of formed bellows and sealed rotary bearings. The joints and bearings are integrated with the composite or pressure retention layer (and its vent network) and with the restraint layer into an assembly called the pressure garment assembly (PGA) (fig. XI-2).

For EVA use, the PGA incorporates a laminated composite for thermal insulation; meteoroid penetration; protection; and resistance to wear, tear, and abrasion. Protective overvisors and coatings provide visual protection from harmful radiant energies. Figure XI-3 illustrates the protective overgarment, the thermal meteoroid garment (TMG), on the pressure suit with the helmet hood and backpack cover removed to illustrate pack control details and to show the helmet visors. The operational lunar suit will have the TMG and PGA integrated as one composite assembly. A suit assembly weight compilation chart is presented in table XI-I.

The prime measure of pressure suit performance is its capability to accommodate the body motions required to accomplish EVA mission tasks. The Apollo soft suit, as presently configured, accommodates the basic performance requirements of the initial Apollo lunar missions, but not without some degradation to the normal shirt-sleeve capability. Joint mobility ranges of the Apollo suit in terms of the percentages of shirtsleeve capability are shown in table XI-II. Although extensive research is being conducted to maximize mobility for all joint areas of the body, it is improbable that full shirt-sleeve capability in a pressurized mode will be achieved. For this reason, it is very important for lunar surface equipment designers to recognize space suit mobility constraints as a trade-off against equipment design. Concerted effort on the part of equipment designers to optimize the astronaut/lunar-surface-equipment interface will be required to fulfill all mission requirements and complete all experiments successfully.

Hard-structure lunar exploration space suit.- For the Apollo Applications Program (AAP) missions in extraterrestrial exploration, an advanced hard-structure space suit is being developed. As the connotation implies, the hard suit consists of a rigid-structure pressure shell connecting each of the mobility joints. The suit was conceived and developed in an effort to provide the following qualities:

1. Added service life
2. A more rugged pressure vessel
3. Increased reliability and mission success

4. Integrated micrometeoroid and abrasion protection
5. A constant-volume, low-torque mobility system

The system consists of (1) a pressure vessel for gas retention, (2) a liquid coolant garment, as in the Apollo soft suit (fig. XI-1), (3) an exterior thermal protective cover layer, (4) insulated gloves and boots for protection from high surface temperatures, and (5) visor filters for eye protection. The suit assembly is shown in figures XI-4 and XI-5. A hard suit assembly weight compilation chart is presented in table XI-III.

The joint mobility range of the Apollo hard suit in terms of the percentages of shirt-sleeve capability is presented in table XI-IV.

Life Support Systems

<u>Portable life support system</u>.- An EVA suit assembly requires support from either an umbilical or from a back-mounted life support package.

The PLSS supplies replenishing oxygen to the circulating pressurized suit internal environment, maintains carbon dioxide and water vapor buildup at acceptable levels, and removes and rejects heat from the circulating fluid heat transport media (water and gas). In addition, the PLSS provides two-way communications capability, which includes generation of audible alarm signals and telemetry of biomedical sensor data. The PLSS also provides an emergency oxygen supply of up to 30 minutes survival time. The PLSS is shown mounted on the suited subject in figure XI-4. Design performance criteria for the PLSS are summarized as follows:

Thermal capacity	
Metabolic	4800 Btu, total 1200 to 1600 Btu/hr, average rates 2000-Btu/hr peaks
External heat leakage	250 Btu/hr in 350 Btu/hr out
Power	4 hours to 240 Wh
Pressure	3.85 ± 0.15 psia 3.2 psia, minimum (emergency)
Carbon dioxide	7.6 mm Hg for the first 2-1/2 hours 10 mm Hg for the next one-half hour 15 mm Hg, maximum
Communications	Redundant two-way Simultaneous voice, one-man EVA Seven channels of telemetry
Rechargeable from spacecraft supplies	Oxygen, water, and LiOH cartridges and batteries
Flow rate of O_2	6 cfm
Weight	70 pounds with 5 minutes emergency time 92 pounds with 30 minutes emergency time
Size	26 inches high, 18 inches wide, and 11 inches thick

Portable environmental control system.- The portable environmental control system (PECS) is an advanced backpack designed to provide gross improvements in space suit support for EVA missions. The major advancement in the design of the PECS is in the use of a solid chemical oxygen supply, rather than of high-pressure gaseous oxygen. Both the primary and emergency oxygen are produced by thermal decomposition of solid chemical sodium chlorate candles.

The design performance criteria for the PECS are summarized as follows:

Thermal capacity	
Metabolic	8000 Btu, total 2000 Btu/hr, average rates 2500 Btu/hr, sustained 3500-Btu/hr peaks for 10 minutes
Environmental	1000 Btu, total 250 Btu/hr in 350 Btu/hr out
Power	To be determined
Suit pressure, outlet	3.80 ± 0.2 psia, normal 3.2 psia, minimum, emergency
Carbon dioxide, maximum inspired PCO_2	7.6 mm Hg, normal 15 mm Hg, spikes and emergency rates
Communications and telemetry	To be determined
Flow rates	6-cfm gas loop 240-lb/hr liquid loop
Modes of operation	Umbilical, O_2, H_2O, communications, and electrical, self-contained
Weight	93 pounds
Size	26 inches high, 16 inches wide, and 6 inches thick
Controls/displays	Located on handheld controller which is stowed on the chest area of the suit

Working in Lunar Environment

As a result of the lunar gravity of one-sixth earth gravity, the solar heat load, the looseness of the lunar soil, and the lack of pressure in lunar environment, the amount of physical work that can be accomplished on the lunar surface in shirt sleeves is much more limited than on earth.

Figure XI-6 indicates the relative time difference in accomplishing various elementary tasks in shirt sleeves, in pressurized and unpressurized suits, and under earth and simulated lunar gravity. Tasks require four to seven times more time than in earth environment, depending on the conditions and the task. The figure also shows that results can be significantly improved by conducting adequate training and by providing proper body restraints, hand and footholds, and so forth. Because of these conditions, every effort must be made by the scientists, principle investigators, and engineers to simplify and minimize the physical work required of the astronaut to increase his chancres of successfully exploring the moon. Tools should be designed to help the astronaut in his activities, and mobility units should be provided to minimize the amount of walking required.

TABLE XI-I.- SPACE SUIT WEIGHT COMPILATION

Component	Weight, lb	Stowed volume, in^3
Liquid coolant garment (filled with 1 lb water)	4.2	
Pressure garment	25.3	5 120
Helmet and visors	6.5	2 744
Thermal cover and boots	13.4	2 160
Total	49.4	10 024 (5.8 ft^3)

TABLE XI-II.- APPROXIMATE PERCENT OF NUDE JOINT

MOBILITY RANGE RETAINED AT 3.7 PSIG

Motion	Apollo soft suit, percent nude range
Upper arm bent outward from side at shoulder	74
Arm bent forward at shoulder	71
Arm bent backward at shoulder	56
Elbow extension	100
Elbow flexion	89
Hand bent inward at wrist (with palm in the horizontal plane)	100
Hand bent outward at wrist	100
Hand bent backward at wrist	90
Hand bent downward at wrist	93
Hip flexion	60
Knee extension	100
Knee flexion	93
Foot bent upward at ankle	88
Foot bent downward at ankle	94

TABLE XI-III.- HARD SUIT ASSEMBLY WEIGHT COMPILATION

[Stowage dimensions: 17 by 26 by 46 inches
Stowed volume: 20 332 in^3 (11.8 ft^3)]

Component	Weight, lb
Liquid coolant garment	3.6
Hard suit with thermal boots	58.7
Helmet and visor	4.3
Thermal cover	6.4
Total	73.0

TABLE XI-IV.- APPROXIMATE PERCENT OF NUDE JOINT

MOBILITY RANGE RETAINED AT 3.7 PSIG

Motion	Apollo hard suit, percent nude range
Upper arm bent outward from side at shoulder	80
Arm bent forward at shoulder	59
Arm bent backward at shoulder	100
Elbow extension	50
Elbow flexion	88
Hand bent inward at wrist (palm in horizontal plane)	100
Hand bent outward at wrist	78
Hand bent backward at wrist	100
Hand bent downward at wrist	69
Hip flexion	70
Knee extension	63
Knee flexion	86
Foot bent upward at ankle	100
Foot bent downward at ankle	100

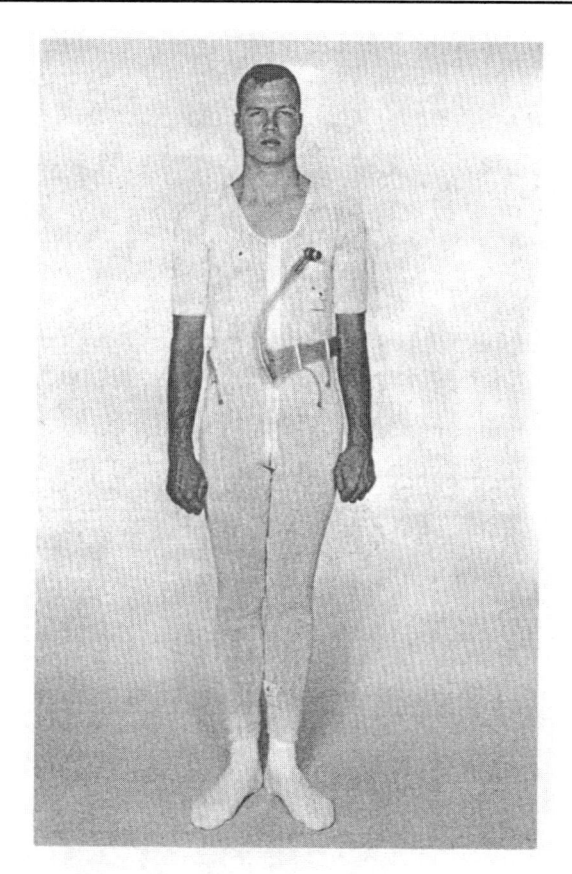

Figure XI-1.- Liquid coolant garment.

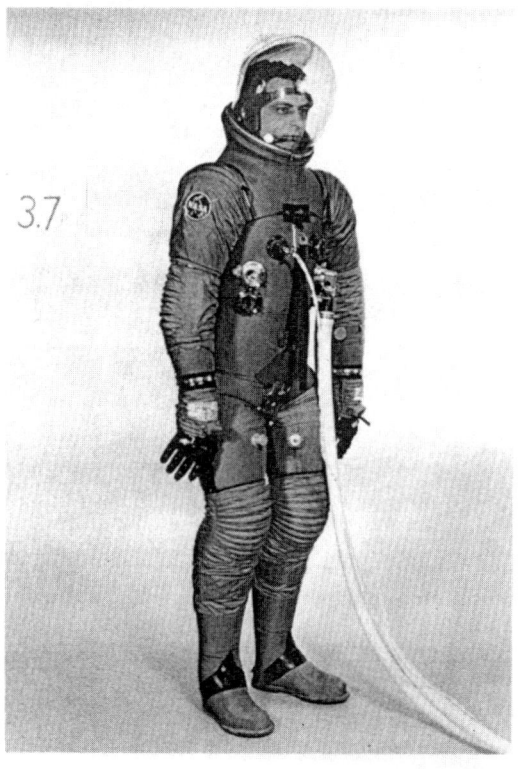

Figure XI-2.- Pressure garment assembly.

Figure XI-3.- Thermal meteoroid garment.

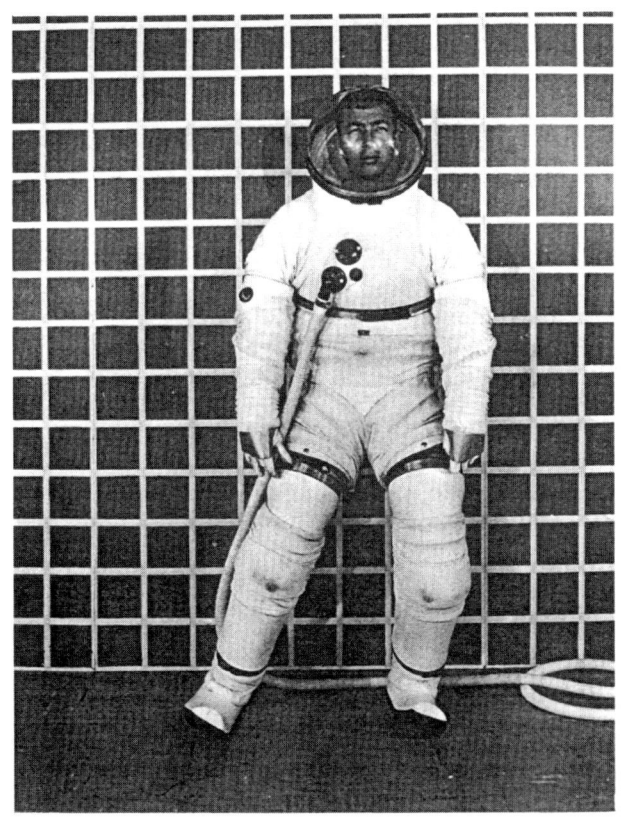

Figure XI-4.- Hard suit, front view.

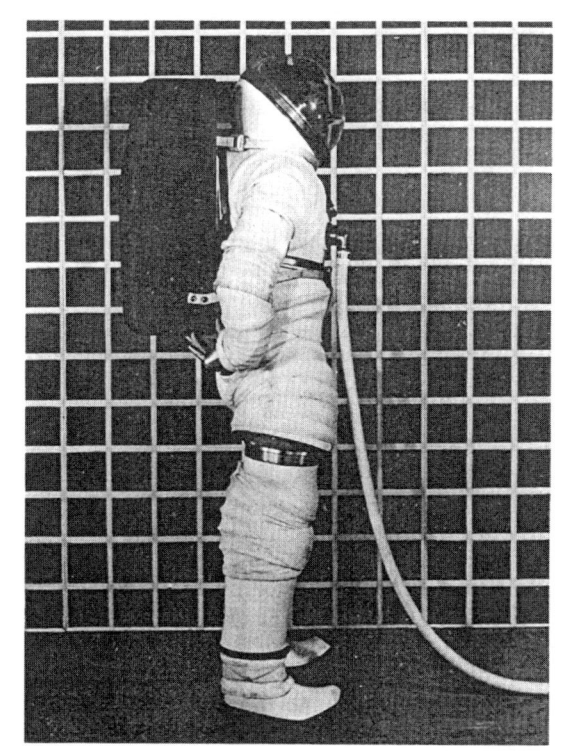

Figure XI-5.- Hard suit, side view showing backpack.

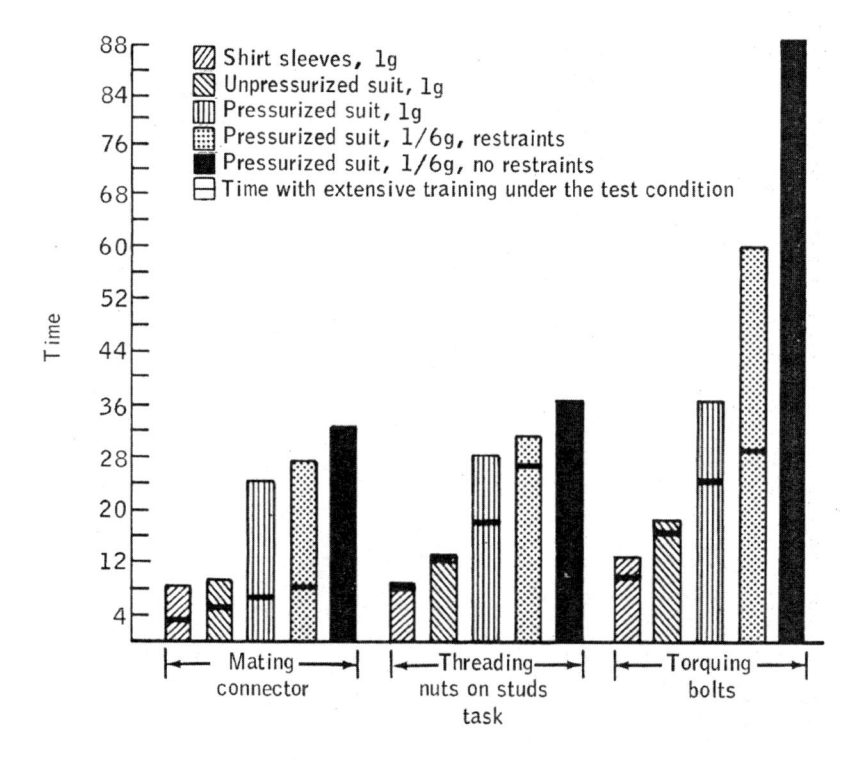

Figure XI-6.- Times to conduct elementary tasks.

SECTION XII

ASTRONAUT LUNAR TIME LINES

The time allocated for scientific investigation during the lunar surface mission must be correlated with the time required for life sustenance of the crew and for essential operations. Planning for lunar exploration and experimentation includes a recognition of the capability and limitations of both man and equipment and insures the high degree of success required of the lunar mission.

The times for the required mission operations are cited in hours per day and are derived from the time lines of various two-man missions of up to 14-day durations.

Time Required for Mission Operations

Nonrecurring operations.- Nonrecurring operations which take place during the first and final days of the lunar stay include the activation and deactivation of the manned lander (requires 2 hours on the first day and 1 hour on the final day of the lunar stay). Examples of nonrecurring operations include aligning the inertial measurement unit (IMU), enabling and disabling the subsystems, and storing the flight equipment. A mission utilizing the lunar module (LM) shelter requires approximately 2 hours for activation of the shelter in addition to approximately 2 hours for deactivation of the manned lander and depressurization for extravehicular activity (EVA). Activation of the LM shelter includes walkaround inspection, airlock deployment, system checkout, shelter pressurization, and the relocation of equipment to an operational configuration. Deactivation of the shelter requires approximately 1 hour, and activation of the manned lander requires approximately 1 hour.

Recurring operations.- Recurring operations require a daily routine for station keeping and maintenance and require approximately 1 hour per day for activities such as the activation of rendezvous radar for command module (CM) track, subsystem checkout, and communications between the lunar crew, the CM, and the Manned Space Flight Network (MSFN).

Time Required for Normal Life Sustenance

The time allowed for eating, relaxation, personal hygiene, recreation, and sleep during an extended mission is approximately 11 hours per day, with simultaneous sleep periods; for both crewmembers, however, missions of shorter duration may allow a shorter and/or staggered sleep schedule.

Time Available for Experiments and Exploration

The time available for lunar scientific activity is distributed in accordance with the requirements of the essential operational time lines. A tabulation of the time lines previously outlined and the available time for experiments and exploration follows.

Activity	First lunar day, hr/day	Lunar stay-time, hr/day	Final lunar day, hr/day
Deactivation of the manned lander	2	--	--
Activation of the LM shelter	2	--	--
Station keeping and maintenance	1	1	1
Personal time	11	11	11
Deactivation of the LM shelter	--	--	1
Reactivation of the manned lander	--	--	1
Total life sustenance and mission operations	16	12	14
Remaining time per day for experimentation and exploration	8	12	10

The time available for lunar experiments and exploration must be utilized in accordance with the type of work to be done, the criteria of human performance, the equipment required, and the safety of the crew.

Intravehicular experiments.- Intravehicular activity experiments performed within the confines of the pressurized manned lander or of the LM shelter, may utilize all the available time. When a depressurized cabin is required, approximately one-third hour is necessary each time the pressure suit is put on and checked out and the cabin depressurized. One-half hour is required to repressurize the cabin and take off the pressure suit.

Extravehicular activity experiments and exploration.- The EVA experiments and exploration must make into account an allowance of approximately 1 hour to don and check out the pressure suit and portable life support system (PLSS), including both egress and ingress of the astronaut. EVA sorties are further restricted by the life support and equipment limitations (section XI). In general, a 3-hour sortie per PLSS (with 1 hour contingency) is available.

Recharge time and battery replacement for the PLSS requires one-fourth hour. Additionally, if a flying-type vehicle is utilized (section X), one-half hour is required for activation of the vehicle. The following table presents an estimate of the time for one sortie using the PLSS and flyer.

Egress

    Don and check out pressure suit   15 min

    Don and check out PLSS   15 min

    Dump cabin and egress to lunar surface   5 min

Exploration and flyer activation (3-hr limit)

    Flyer activation   30 min

    Exploration   150 min

Ingress

    Ingress to vehicle and hookup to LM ECS   10 min

    Repressurize vehicle and doff pressure suit   20 min

                                                                 4 hr 5 min

SECTION XIII

EXPERIMENT MODULES

Apollo Lunar Surface Experiments Package

The Apollo lunar surface experiments package (ALSEP) is designed to collect and transmit lunar scientific data to the earth for up to 1 year after the departure of the astronauts. The ALSEP is scheduled to be included on the first three Apollo manned lunar landings.

ALSEP system objectives will be achieved by seven scientific experiment instruments and the related supporting subsystems. The astronaut will arrange a group of the experiment instruments and related subsystems on the lunar surface approximately 300 feet from the lunar module (LM). He will also set up sensors and then start the package operation. While in operation on the moon, the ALSEP system will be self-sufficient, using a radioisotope thermoelectric generator (RTG) for electrical power. Command and data communications will be established and controlled by way of the Manned Space Flight Network (MSFN). ALSEP commands will originate within the Mission Control Center, Houston (MCC-H), and will be forwarded to remote sites of the MSFN. At the same sites, telemetry data received from the ALSEP will be forwarded to the MCC-H.

The ALSEP telemetry system (fig. XIII-1) consists of two distinct links: an earth-to-moon command link and a moon-to-earth scientific and engineering data link. The ALSEP earth-to-moon link (the up-link) provides for the remote control of ALSEP command functions such as experiment mode selection, transmitter selection, change of subsystem data rates, and subsystem operation flexibilities (turnon, turnoff, etc.). The ALSEP moon-to-earth link (the down-link) provides for the transmission of scientific and engineering data from the ALSEP subsystems to earth receiving stations. Three data transmission frequencies will be used to permit the simultaneous operation of up to three separate ALSEP systems deployed and operating on the lunar surface. The bit rates are 10 600, 1060 (normal), and 530 bits/sec. The basic data format is an 8-by-8 matrix of 10-bit words.

The electrical power subsystem provides a minimum of 46 watts of power at +16 V dc (nominal). The total ALSEP package (including supporting subsystems) occupies a volume of approximately 15 cubic feet and weighs approximately 255 pounds. The storage location on the LM is shown in figure XIII-2.

The ALSEP is deployed by the astronaut in a prescribed sequence. Each experiment instrument is connected to a central station by a flat, ribbon-like conductor. The central station consists of the data subsystem, electronics for the seismic instruments, and a control panel for activation of the system by the astronaut. The central station is the focal point for the interconnection of the ALSEP subsystems and is the focal point for the transmitting and receiving of data to and from the MSFN.

ALSEP Experiments

The following eight experiments have been selected by the National Aeronautics and Space Administration (NASA) as part of the ALSEP system.

1. Active seismic experiment (ASE)
2. Charged-particle lunar environment experiment (CPLEE)
3. Cold cathode gage experiment (CCGE)
4. Heat flow experiment (HFE)
5. Lunar surface magnetometer (LSM)
6. Passive seismic experiment (PSE)
7. Solar-wind spectrometer (SWS)
8. Suprathermal ion detector experiment (SIDE)

Because of various constraints on the ALSEP system, a smaller number of combinations of the experiments are scheduled for each ALSEP mission. Typical ALSEP systems are shown in figures XIII-3 and XIII-4. At present, the selected experiments are:

1. ALSEP system 1; CCGE, LSM, PSE, SWS, SIDE
2. ALSEP system 2; CCGE, LSM, PSE, SWS, SIDE
3. ALSEP system 3; CCGE„ CPLEE, HFE, PSE
4. ALSEP system 4; ASE, CCGE, CPLEE, PSE, SIDE Lunar Surface Drill

Lunar Surface Drill

An early Apollo mission will be equipped with a self-contained lunar drill to bore two holes, 10 feet deep and 1 inch in diameter, in which two heat flow experiment probes are emplaced. The capability of the drill, limited by battery life, is sized for the following lunar subsurface models:

I. A depth of 10 feet in pumice or softer material
2. A depth of 4.1 feet in 43 percent vesicular basalt
3. A depth of 10 feet in unsorted, uncohesive conglomerate (including one-half foot of dense basalt)
4. A depth of 10 feet in unsorted, uncohesive conglomerate (including 2-1/2 feet of vesicular basalt)
5. A depth of 0.75 foot in dense basalt

The drilling rate varies from 1 to 60 in/min depending on the lunar surface. These drilling rates are based on a drill load of 6 to 10 pounds. Sufficient casing is provided to maintain hole integrity. The core of the second hole will be returned for geological examination. The drill can also be used to obtain samples from outcroppings.

The storage volume of the drill is approximately 1 cubic foot. The drill is powered by a nominal 23-V silver-zinc battery with a drill load of 6 to 10 pounds. The battery is rated at 300 Wh with a maximum discharge rate of 18.75 amperes. The battery is 4 by 4 by 3 inches in size and weighs 7.5 pounds.

Deep Drill for Lunar Surface

A study of the deep drill is in process. Tentative specifications require a 2-inch-diameter hole up to 100 feet in depth. Such a drill will weigh approximately 250 pounds in contrast to the 27-pound drill carried on the present Apollo. This drill must be mounted on the LM or some heavy vehicle. Percussion and rotary drills are being considered.

Percussion drill.- The percussion drill system is comprised of the following major systems:

1. Pneumatic system
2. Heat rejection system
3. Instrumentation and control system
4. Low pressure plenum and turret section
5. Flexible drill string drum and drive
6. Downhole drill system

Rotary drill.- The rotary drill system is comprised of the following major components:

I. Drill frame
2. Sealed drive motor
3. Gearbox
4. Drill string
5. Bit coolant unit
6. Core recovery unit
7. Control and protection unit

Electrical power requirements.- The electrical power requirements are as follows:

Percussion	Power
Compressor	28 V dc, 4 horsepower (3kW)
Reel drive motor	28 V dc, 0.5 horsepower
Torque motor	28 V dc, 3 amperes
Rotary	
Drive motor	100 V dc
Sealed model	115 V ac, 60 cycle
Air-cooled model	28 V dc
Controls	28 V dc (filtered)

Schedule and funding.- Approximately 45 months will be required to complete the phase D effort with an expected total cost of $19.18 million.

Lunar Survey System Staff Device

The lunar survey system staff device resembles a staff and is handheld by the astronaut for use on the lunar surface (fig. XIII-5). The staff device is used in conjunction with a laser tracker which logs the staff position in relation to the tracker. The following are objectives of the device.

1. Field measurements of the lunar surface
2. Determination of positions of samples and data points
3. Construction of a geometric array from photographic images

The staff device contains the following:

1. Television (TV) camera subsystem
2. Stereophotogrammetric camera subsystem
3. Digital telemetry subsystem
4. Staff orientation subsystem
5. Self-contained power supply

The device can be carried by the astronaut during each traverse over the lunar surface to a maximum range of 700 meters from the LM.

Television camera subsystem.- A shuttered slow-scan vidicon TV subsystem used with a modified Apollo TV camera takes an 800-line picture every 3.2 seconds.

Stereophotogrammetric camera subsystem.- A stereophotogrammetric camera is employed to make precise, matched, stereophotographs of random lunar surface features for photogrammetric interpretation. To provide more versatility, the camera is detachable from the lunar survey system staff device. The camera takes 24- by 36-mm pictures on 35-mm film. The stereo lenses have a 35-mm focal length with a 150-mm stereo base. In addition to stereo pictures, the camera simultaneously takes a telephoto picture with a 135-mm-focal-length lens. The exposure is recorded so that photometric data can be recovered from the negatives. Filters are provided to permit polarized pictures at three different polarizations; color separation filters are also provided. The significant parameters such as focus distance, exposure duration, filter in use, and time of exposure are recorded on the film.

Digital telemetry subsystem.- A digital telemetry subsystem correlates the staff device position (tip, tilt, and azimuth) with the transmitted TV images.

Staff orientation subsystem.- The continual monitoring of the following six spatial parameters is essential to the accomplishment of the objectives of the staff device.

1. Range (l0 to 700 meters, maximum)
2. Elevation
3. Azimuth
4. Tip
5. Tilt
6. Staff pointing azimuth

The first three parameters are monitored by a laser tracker. The device automatically tracks and ranges the staff. The device senses a laser energy return from a retroreflector mounted on either the staff or the astronaut. The laser energy return provides very accurate angular measurements. The range is derived from transit-time measurements to an accuracy of 10.5

meter. The staff orientation parameters (tip, tilt, and pointing azimuth) are measured by linear pendulous accelerometers mounted on the staff to an accuracy of 0.06°. The six parameters give the position, size, shape, and orientation of features seen by the camera and analyzed by photogrammetric techniques.

Self-contained power supply.- Silver-zinc secondary batteries appear to be the most suitable as a power supply for the lunar survey system staff device.

Radiofrequency communications link.- A radiofrequency communications link between the staff device and the LM carries TV and telemetry (and possibly an added backup voice) capability.

Weight and power summary.- The weights of individual equipments within the staff and their power requirements are given in the following table. The total storage volume is 12 cubic feet.

Schedule and funding.- The cost to produce one field-test article scheduled for completion in mid-1968 will be approximately $1 million.

Item	Weight, lb	Power, W
Instrumentation	3.1	6.0
Telemetry system	0.5	1.5
Transmitter	1.0	0.6
Ranging/tracker	19.5	22.2
TV system	7.0	8.0
Mechanical	7.0	
Battery	25.2	
Camera	9.92	5.0
Receiver/repeater	5.0	55.0
Totals	78.22	98.3

Lunar Mapping and Survey System

The lunar mapping and survey system (LM&SS) is a high-performance camera system carried in a module located in the Apollo-Saturn adapter. The LM&SS replaces the LM on certain lunar missions and is operated in lunar orbit in a docked configuration. The camera system is composed of two instruments: a high-resolution survey camera and a moderate resolution mapping camera in combination with a stellar index camera.

The film is retrieved from the LM&SS by the astronaut at the completion of the mapping and survey mission. Approximately 75 pounds of film is recovered from the camera systems. Four LM&SS articles are being developed for the Apollo Applications Program (AAP) missions.

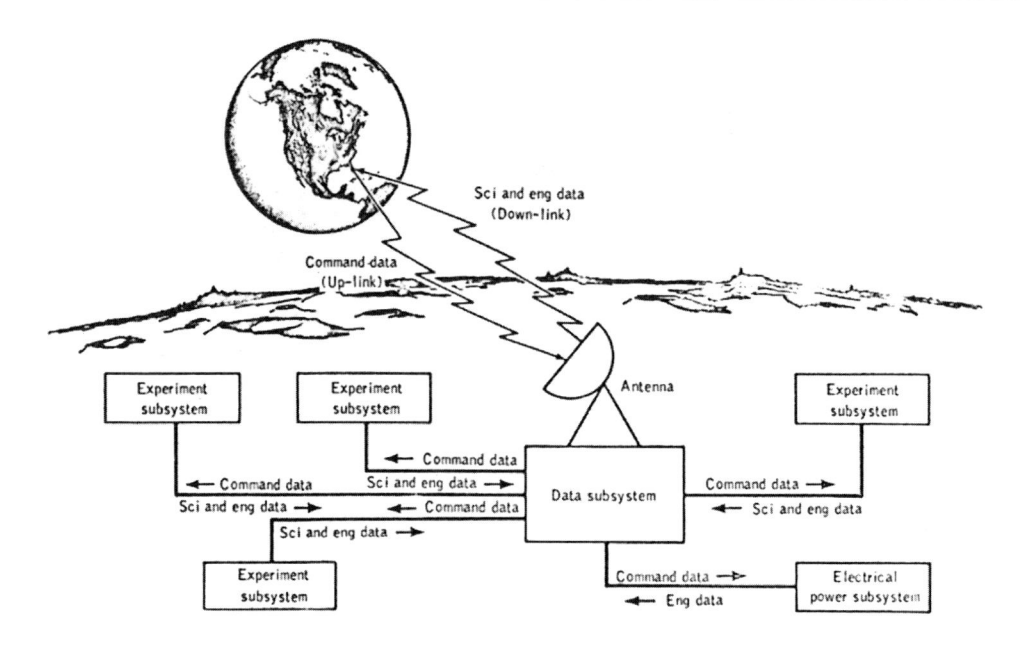

Figure XIII-1.- Apollo lunar surface experiments package telemetry system.

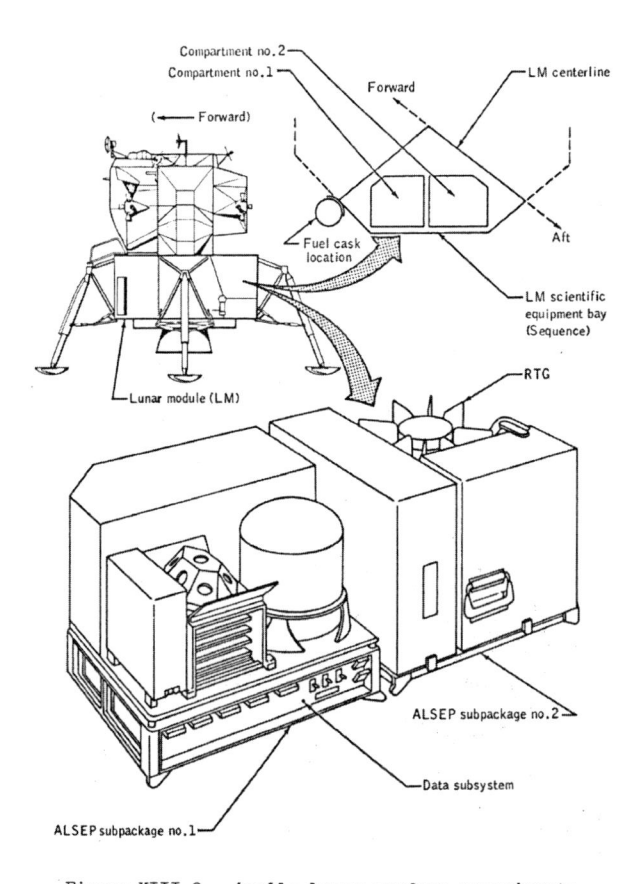

Figure XIII-2.- Apollo lunar surface experiments package delivery configuration.

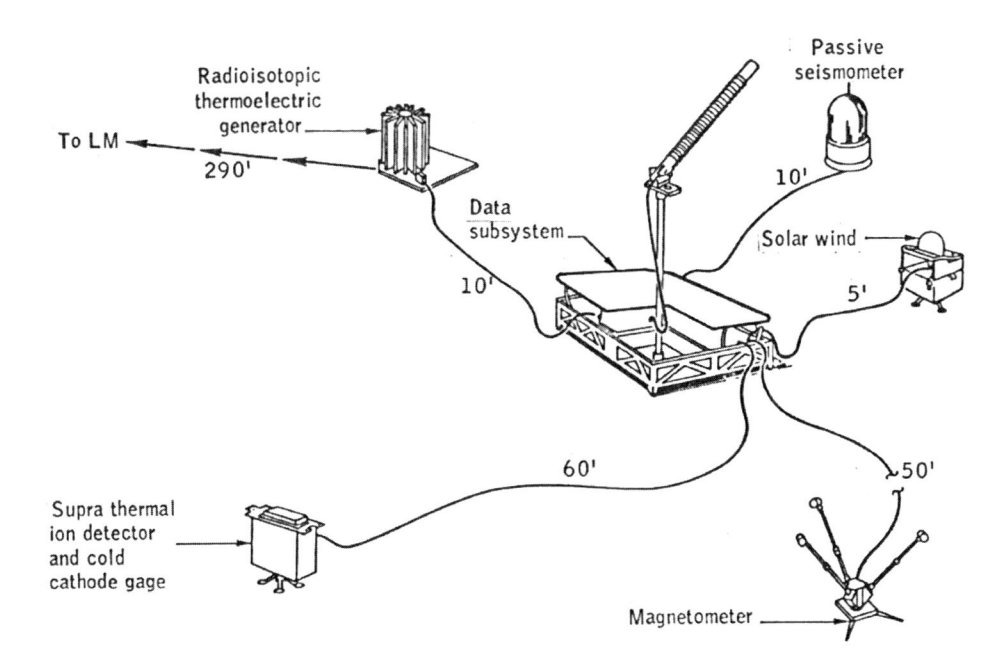

Figure XIII-3.- Apollo lunar surface experiments package array A (typical).

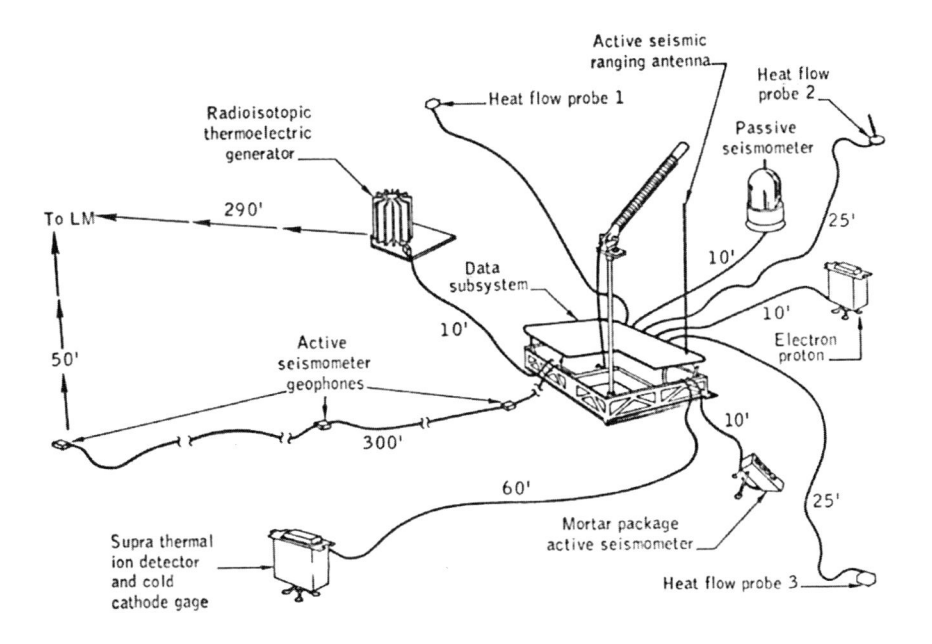

Figure XIII-4.- Apollo lunar surface experiments package array B (typical).

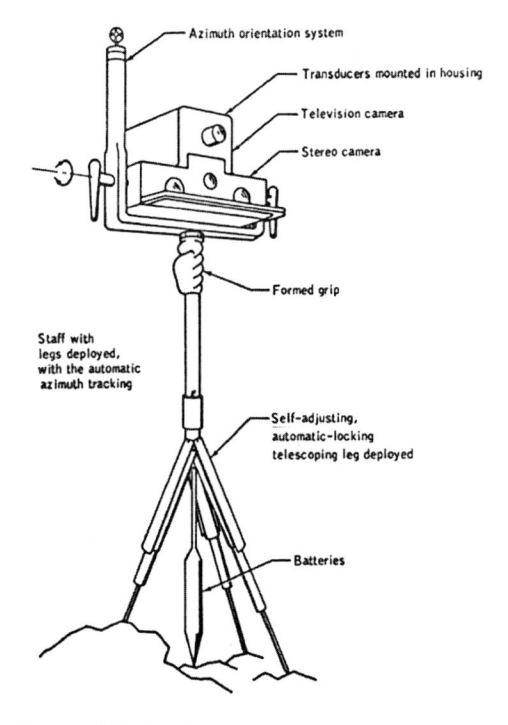

Figure XIII-5.- Lunar survey system staff.

SECTION XIV

EXPERIMENT EQUIPMENT REQUIREMENTS

The gathering of meaningful data from scientific investigation of the moon is predicated on an experiments schedule developed early in the mission planning. This schedule implements a plan which. (1) defines the experiment and vehicle interface, (2) develops the hardware and support equipment, and (3) demonstrates the operation and compatibility of the experiment to be within the framework of a mission profile. The restrictions imposed upon an experiment, as defined by such schedules, establish specific requirements concerning availability, qualifications, and operation of the experiment equipment. These requirements, as outlined below, furnish the time lines and milestones essential to achieving early lunar exploration objectives.

Equipment Availability Requirements

The availability of experiment equipment must be compatible with the mission schedule. Based upon the experience gained from prior manned spacecraft programs, requirements have been established for both leadtime and hardware availability.

Mass mockup hardware.- The mass mockup must be an exact model with the correct external dimensional configuration and with the weight, the center of gravity, the functional controls, the electrical connections, and the mounting provisions of the flight hardware. The mass mockup need not be a functional model, but the method of operation and any critical axis-mounting directions must be indicated. One unit is generally required with delivery 18 months prior to launch.

Qualification test hardware, flight hardware, and flight backup hardware must be identical in configuration and production processing. The qualifications testing is required to be completed several months prior to delivery of the flight and backup hardware. In general, this hardware must be delivered 12 months before launch. More than one unit of qualification test hardware may be required depending upon individual experiments, test requirements, and schedules.

Training hardware.- The training hardware must be identical to the flight configuration, but need not be flight qualified. Training hardware must be delivered not later than the flight hardware. Depending on the individual experiment, one or more units may be required.

Ground support equipment.- Ground support equipment (GSE) which is peculiar to the requirements of an experiment and not readily available as Government-furnished equipment (GFE) must be available not later than the flight hardware. Only one unit is generally required, but the possibility exists, dependent upon the individual experiment, of a requirement for one unit at the vehicle assembly site and one unit at the launch site. Repairability and availability of spare parts should be considered.

Equipment Qualification Requirements

This section provides a guide for flight qualification of experiment equipment with respect to the environments expected to be encountered during a mission. The equipment tested must be identical in configuration and production processing to the flight equipment and generally will not be used as flight hardware.

The sequence of tests normally must follow the same order in which the environments will be encountered during the mission. Functional tests, to determine whether the hardware is performing within specification tolerances, must be conducted before and after each environmental exposure. The same functional test must be performed during the environmental exposure if the equipment is required to operate in that environment during the mission.

The following are environmental tests required for qualification of experiment equipment.

l. The qualification test requirements for the earth environment include humidity, salt fog, high and low temperature, shock, vibration, and fungus (ref. 2).
2. The qualification test requirements for the spacecraft environment include pressure, acceleration, vibration, noise, oxygen compatibility (ref. 1), electromagnetic interference, magnetic fields, and electrical compatibility (ref. 3).
3. Space environment qualification test requirements for the space environment include low pressure, solar energy, and meteoroid impact (ref. 2).

Training Requirements

Training requirements vary with the complexity of each experiment. Those which require extensive interface with the vehicle or are hard mounted to the vehicle must provide training hardware for each simulator and for one backup trainer. That equipment which can be classified as portable should provide one trainer and one backup unit. It is also expedient to provide a training unit at the vehicle contractor's facility for interface verification and system testing. As previously indicated, training hardware must be available 12 months prior to launch.

In addition to the training hardware requirements, certain documentation is necessary to support the experiment. Requirements which involve the flight crew before, during, and after the flight must be specified. Checkout procedures must be furnished for any special requirements involving communications, telemetry, recovery, and data handling.

The experimenter is required to brief the flight crew on the history and purpose of the experiment and must be readily available for consultation in the event flight plan deviations are necessary during a mission.

The following is a typical example of an experiment-development schedule indicating the hardware end-item quantities and the required delivery schedule:

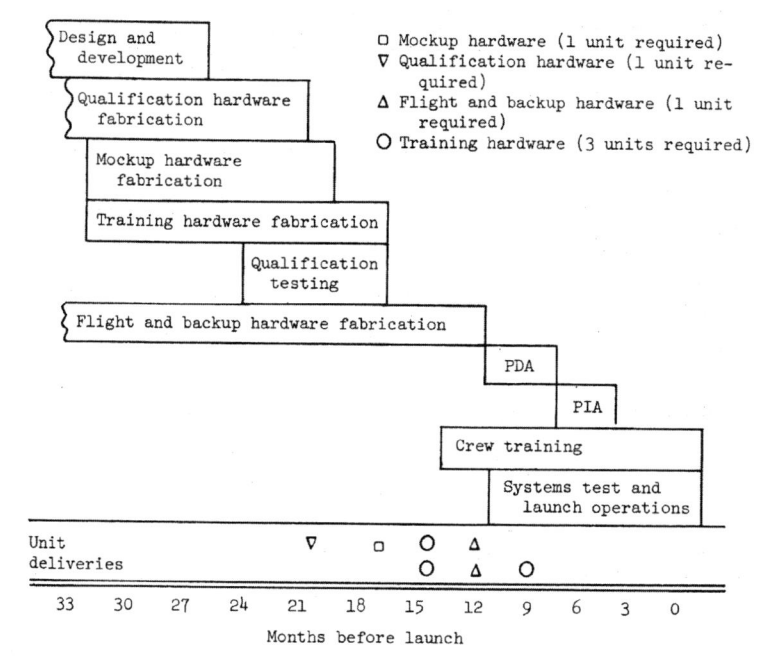

```
Design and development
Qualification hardware fabrication
Mockup hardware fabrication
Training hardware fabrication
Qualification testing
Flight and backup hardware fabrication
                                    PDA
                                        PIA
                              Crew training
                                  Systems test and
                                  launch operations

□ Mockup hardware (1 unit required)
▽ Qualification hardware (1 unit required)
△ Flight and backup hardware (1 unit required)
○ Training hardware (3 units required)

Unit deliveries       ▽    □  ○  △
                                ○  △   ○
         33   30   27   24   21   18   15   12   9   6   3   0
                          Months before launch
```

SECTION XV

LAUNCH VEHICLE CHARACTERISTICS

This section presents a brief description of the Saturn V (S-V) launch vehicle and some of the launch operational constraints. Figure XV-1 is an illustration of the S-V launch vehicle and the Apollo spacecraft. The S-V has three propulsion stages and one instrument unit. The S-IC stage develops about 7.5 million pounds of thrust by burning liquid oxygen and RP-1 in the five F-1 engines. Liquid oxygen and liquid hydrogen are burned in the five J-2 engines of the S-II stage to provide about 1 million pounds of thrust. The S-IVB stage uses liquid oxygen and liquid hydrogen in its single J-2 engine (200 000 pounds of thrust) to insert the spacecraft into earth orbit. The S-IVB stage is reignited to provide propulsion for translunar injection from earth orbit.

Payload Capability

The S-V has the capability to inject about 98 000 pounds onto a 72-hour flight from earth orbit to the moon. Small increases in this payload can be obtained by the following means:

	Approximate translunar payload increase, lb
Increase translunar flight time to 110 hours (nonfree-return trajectories)	4900
Perform translunar injection during first earth orbit	1500
Direct translunar injection (may not be practical)	6000

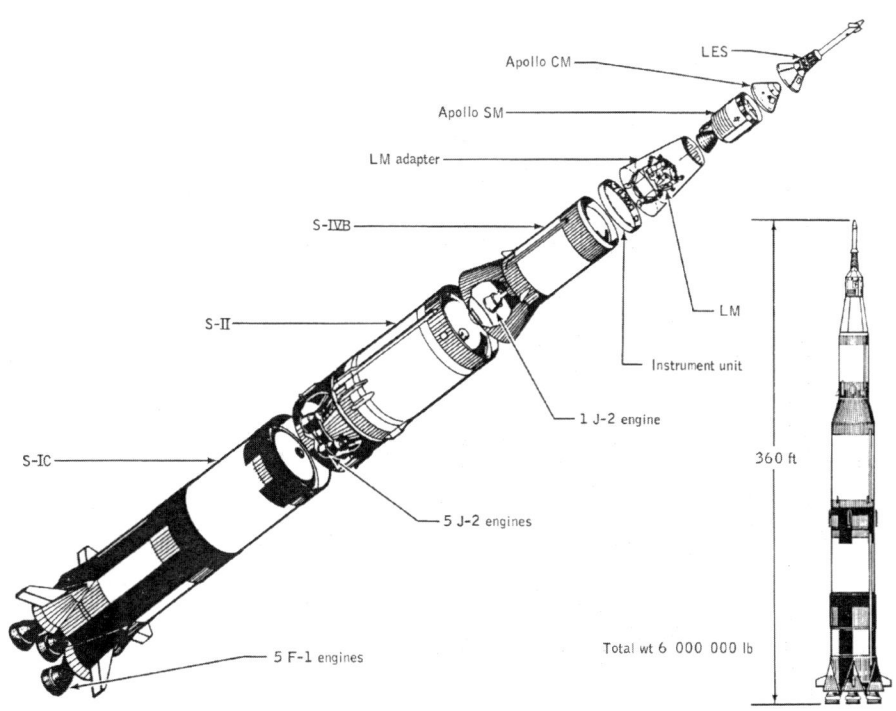

Figure XV-1.- Saturn V launch vehicle and Apollo spacecraft.

The indicated payload increases are not necessarily accumulative. For example, direct injection and first-orbit injection cannot both be done. Also, free-return trajectories and translunar flight time are interrelated. The payload increases shown are each associated with only the single change indicated.

These payload increases (approximations obtained without benefit of a rigorous analysis) are presented to indicate some potential means for obtaining small translunar payload increases. This payload includes the scientific equipment and spacecraft propellant necessary to deliver that equipment. These operational changes to increase the payloads may or may not be achieved practically. Hopefully, as more experience is obtained and more confidence is established, these payload increases may be available.

Modifications of the F-1 thrusting magnitude-time sequence and the replacement of the J-2 engines with the J-2S (an uprated version) engines can also improve the translunar injection payload capability of the S-V launch vehicle. Increased payloads as high as 10 000 additional pounds have been estimated.

Translunar payloads can be increased greatly, of course, by major launch vehicle changes, such as by strapping on large solid-propellant motors, by uprating engines, and by adding propellant. Payloads as high as 234 000 pounds have been indicated. Because of the long leadtime necessary for major launch vehicle changes, these large upratings will probably not be available for early lunar exploration.

The weight above the launch vehicle instrument unit (and below the launch escape system) that the launch vehicle can successfully launch without structural failure is termed the stack limit. The stack limit places a restriction on the launch vehicle payload. The payload must weigh no more than the allowable stack limit. Payloads of 100 000 pounds have been investigated without exceeding the standard S-V stack limit. Stack limit analyses above 100 000 pounds have not been made to date, but the launch vehicle inherently may have considerable capacity beyond this point.

Kennedy Spacecraft Center Launch Operations

A very brief outline of the procedure and time required for launch operation is presented in figure XV-2. About 4-1/2 months before launch the spacecraft is required at Kennedy Spacecraft Center (KSC). The lunar module (LM) ascent and descent stages are received, inspected, and checked in the Manned Spacecraft Operations Building (MSOB), including a manned altitude run and a docking check with the command module (CM). The command and service modules (CSM) are received, inspected, and checked in the MSOB, including a manned altitude run. Spacecraft test flow through the Vertical Assembly Building (VAB) and on the launch pad follows procedures developed for the nominal Apollo program. Figure XV-2 is based on the assumption that all experiments are installed in the spacecraft before initiation of the vehicle flow through KSC assembly and checkout procedures. The experiments are installed 4-1/2 months before launch. The spacecraft are assembled and physically mated to the launch vehicle approximately 2-1/2 months prior to launch.

The launch vehicle stages and the instrument unit are received about 3 months before launch, then inspected, and erected in the VAB. Checkout is performed in the VAB with the launch vehicle erected on the mobile launcher. Individual stage testing is completed prior to overall tests. The entire space vehicle is assembled on the mobile launcher in one of the VAB high bays; then, the vehicle is transported to the launch complex by a crawler transporter (CT) about 1 month before launch. Space vehicle flight readiness is established by the simulated flight test (SFT) in the VAB and by a flight readiness test (FRT) at the launch area. After the mobile service structure (MSS) is removed from the launch area during the countdown, servicing of the spacecraft is no longer possible. Removal of the MSS is initiated about 9 hours before launch.

Lunar mission countdown and hold-decision points for the S-V are shown in figure XV-3. There are several critical points in the space vehicle countdown that determine the hold time available. The decision to hold at these various points could affect the availability to launch.

Launch Vehicle Characteristics

Countdown time, hr	Consideration
To T-37	Status can remain fairly constant. Safe and arm devices are required to be validated 120 hours prior to launch.
T-36.5	Spacecraft is serviced with gaseous helium.
T-30	Fuel cells are activated. Further holds decrease mission life.
T-20	Spacecraft fuel cell liquid oxygen loading is completed.
T-17	Fuel cell liquid hydrogen loading is completed. Fuel cell consumables available are decreased by time in hold.
T-15	LM super-critical helium top-off is completed. Present limitation for descent engine ignition to occur is within 130 hours thereafter.
T-9.5	Mobile service structure is removed.
T-8	Launch vehicle cryogenic loading begins.
T-4	Launch vehicle cryogenic loading is completed. Approximately 8-hour hold time is available with present cryogenic storage.
T-3	Crew ingress occurs.
T-6 min	Safe and arm units are armed.

The time required to turnaround (recycle for another launch attempt after a major hold) from various points in the launch countdown (fig. XV-4) has effects which are closely related to those considered for holds. The turnaround times are increased after servicing lines are disconnected, after hypergolic systems are wetted, and after propellant systems are loaded. A scrub close to T-0 requires approximately 44 hours of activity to reach a new T-0. A scrub prior to T-10 hours would enable a launch attempt the following day; but, a scrub later than that time would require a 2-day turnaround time. Readjustment of the spacecraft service propulsion system (SPS) propellant loads would require approximately 1 additional day.

When the countdown is rescheduled in its entirety for some future time, there are certain considerations that must be made. Space vehicle problems must be corrected prior to a new launch attempt, and there are constraints due to lifetime limitations on flight hardware. Figure XV-5 shows a typical reschedule from one lunar month to the next. It is assumed that the S-IVB auxiliary propulsion system (APS) units will be changed, since the units are qualified for exposure to propellants for only 30 days. However, the exposure to propellants possibly will be changed in the future to 90 days, thus eliminating that activity. The spacecraft propellant system presumably will have been qualified for a longer lifetime than at present; otherwise, the attempt in the second month will not be possible. The fuel cells probably will not be replaced because a fuel cell replacement on the pad will require approximately 19 days, making it impossible to meet launch opportunities for the second month after having missed those opportunities in the first month. In the turnaround plan shown, no countdown demonstration will be required prior to the launch attempt in the next month, and the vehicle will not be returned to the VAB from the pad. Probably more than 2 weeks will be required to return the vehicle to the VAB and subsequently back to the pad (in the event of a hurricane or some other cause for move which occurs late in the checkout testing operations).

Lunar missions requiring dual launches, such as a taxi and shelter mission, impact the launch considerations in two ways. First is the utilization of KSC facilities, and second (and closely related) is the minimum time between two S-V launches when both KSC launch pads 39A and 39B are used. The sequence is based on the present Apollo mission procedures.

These procedures may require less time as more experience in S-V launches is obtained. Current best estimates for minimum time between launches is 24 days. No MSS changes are required between launches (fig. XV-6), such as those changes which may be necessary if different spacecraft and/or scientific payloads are flown on the two launch vehicles.

Dual-launch missions would, of course, cause a higher-density load on the KSC facilities. In addition to the VAB and pad facilities, the MSOB, with its two power LM stands and its one power CSM stand and five acceptance checkout equipment (ACE) stations, are affected. These facilities also are to be used by other nonlunar missions.

Activities for the CSM and LM assigned to a dual-launch mission are similar to those for the single-launch mission. The MSOB is not capable of supporting simultaneous checkout of two complete spacecraft (two command and service modules and two lunar modules). With proper scheduling, it is anticipated that checkout; can be performed with minimum delay based upon work performed and reported in reference 1.

If spacecraft assigned to other missions are scheduled for checkout operations in the MSOB during the same timespan, work areas (such as the altitude chamber) may be overloaded. Therefore, operations supporting the spacecraft for different missions may require the establishment of a priority schedule.

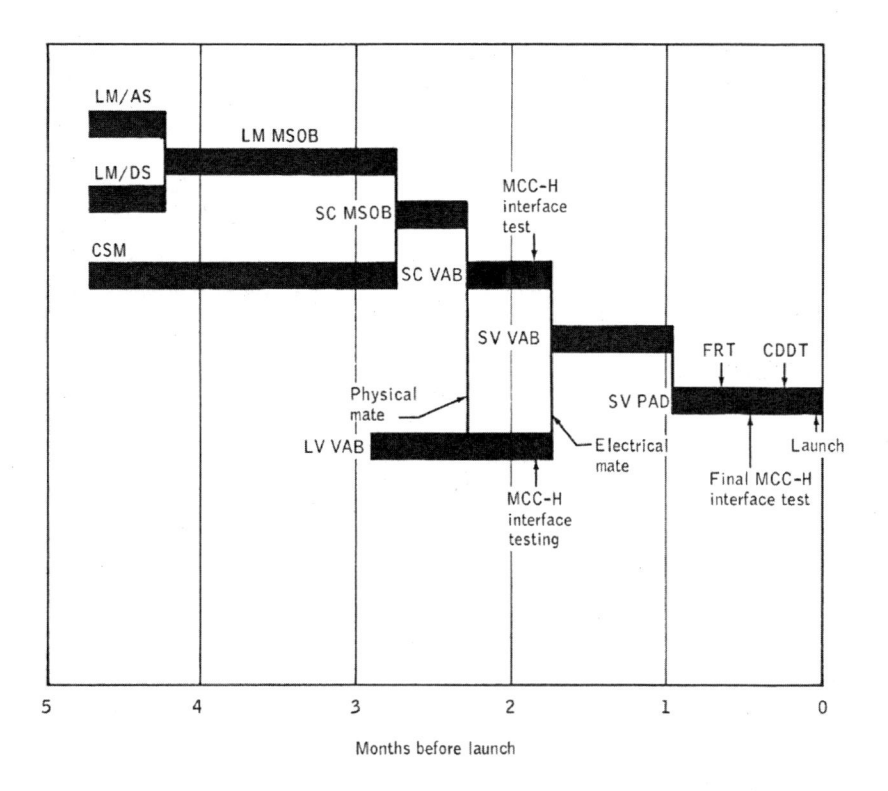

Figure XV-2.- Space vehicle flow.

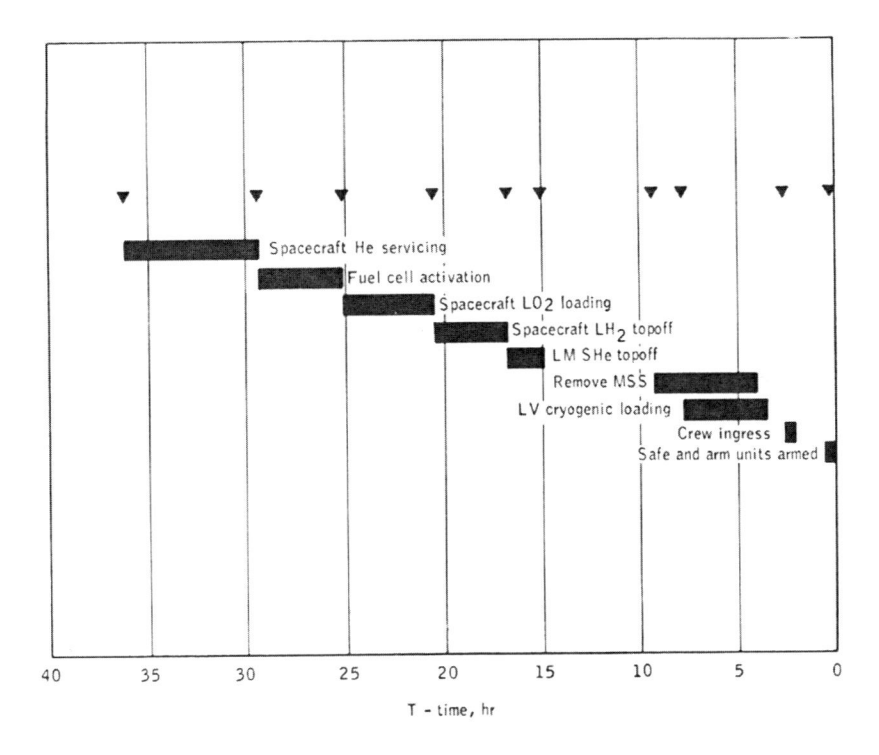

Figure XV-3.- Saturn V countdown and hold decision points.

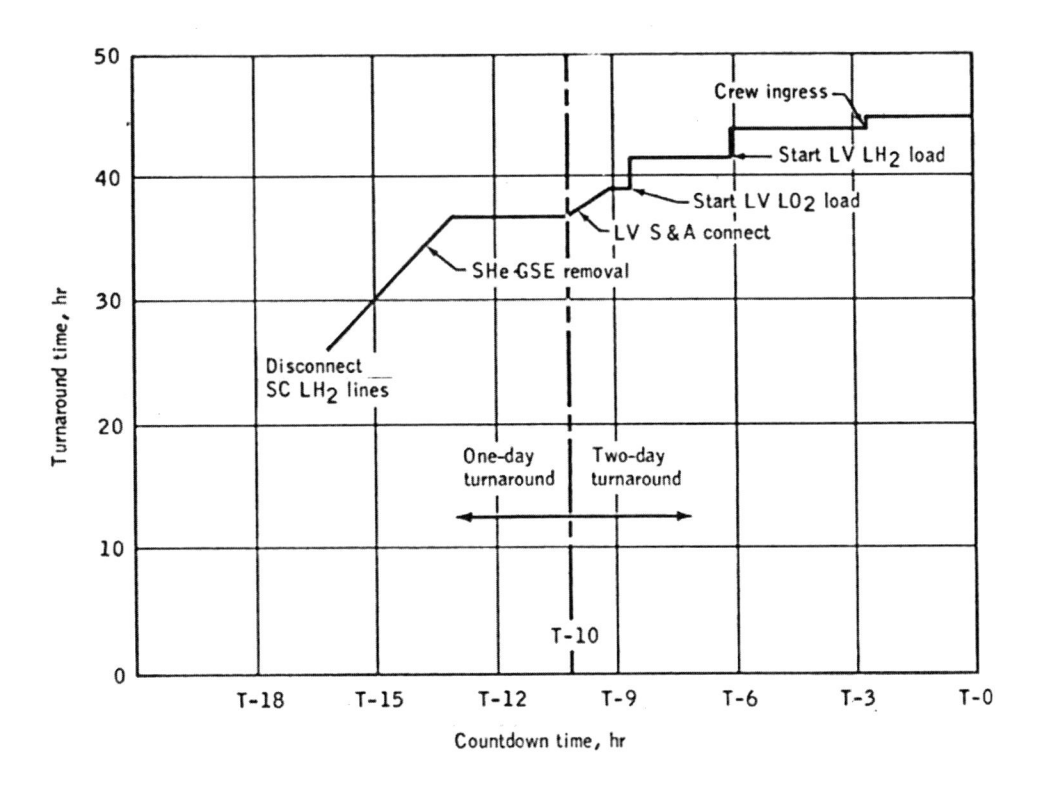

Figure XV-4.- Saturn V turnaround time from scrub to next T-0.

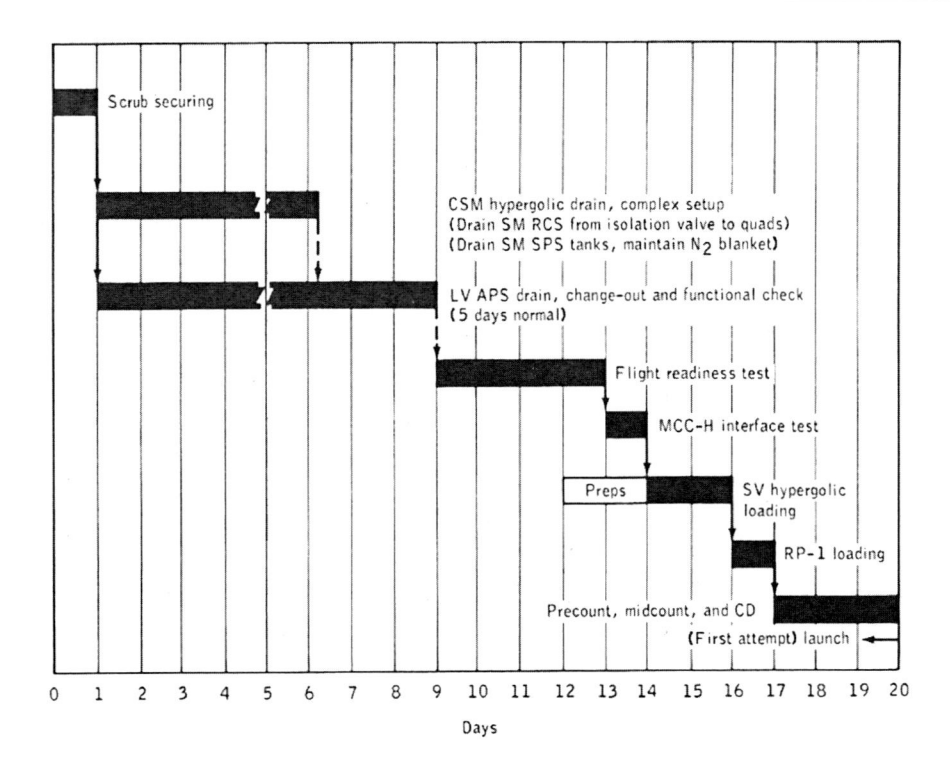

Figure XV-5.- Space vehicle reschedule plan.

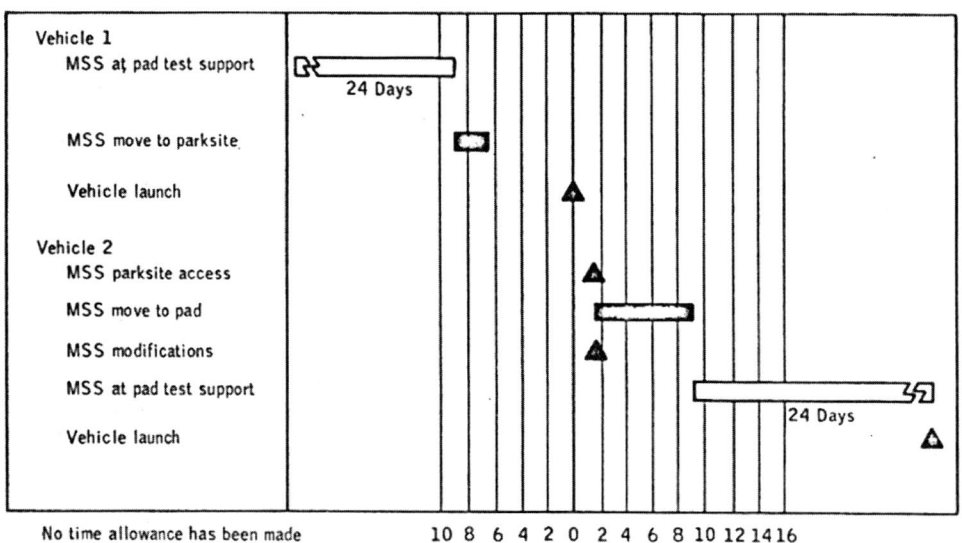

Figure XV-6.- Mobile service structure flow plan, two-vehicle launch.

SECTION XVI

APOLLO COMMAND AND SERVICE MODULE

The Apollo spacecraft, including the Apollo command and service module (CSM), is the basic vehicle used to transport the crew to lunar orbit for transport to the lunar surface. Figure XVI-1 shows the spacecraft in the launch configuration with the lunar module (LM) stored in the spacecraft lunar module adapter (SLA).

Command Module

The command module (CM) has the facilities for housing the three man flight crew and contains the necessary automatic and manual equipment to control and monitor spacecraft systems. The CM also contains the required equipment for the safety and comfort of the flight crew and contains the earth landing system for parachute recovery of the spacecraft after reentry into the atmosphere of the earth. The CM weighs approximately 13 000 pounds. The crew compartment has a volume of approximately 245 cubic feet and a total pressurized volume of 306 feet. The return payload capability is presently 100 pounds, but can probably be increased for future missions. Payload return is limited by volume more than by weight.

Service Module

The service module (SM) is a cylindrical structure which contains the main propulsion system and the reaction control system that are used for midcourse corrections, lunar orbit insertion, and transearth injection. This propulsion system is also used for lunar orbit plane changes. The SM houses the fuel cells for electrical power and most of the onboard consumables for the spacecraft.

Bay 1 of the SM is a 50° sector containing approximately 170 cubic feet of usable volume which can be made available for scientific equipment on certain missions. This volume, with reference to the command module, is located aft and to the right of the spacecraft access hatch. This can be seen in figure V-1 in section V. The SM weighs approximately 11 000 pounds and carries approximately 40 000 pounds of propellant in its tanks.

Uprating the SM would be a major task. The propellant tanks are the length of the SM and are essentially full for the present missions, using a 98 000-pound-capacity booster. Uprating the SM would require increasing the size of the tanks, which would require increasing the size of the SM. A change of this nature would require a major modification to the launch-facility service tower and structural changes in the Saturn V (S-V) launch vehicle and in the SLA to accommodate the added weight.

Spacecraft Lunar Module Adapter

The SLA is a truncated cone which connects the CSM to the S-IVB stage of the launch vehicle. The SLA is 154 inches in diameter at the CM interface and 260 inches in diameter at the aft-end interface. The SLA is constructed of eight panels. The four forward panels are movable, and, at separation of the spacecraft from the SLA, these panels rotate outward 45° from the vertical, petal fashion, giving easy access for the CSM to dock and extract the LM which is housed in the SLA from launch through translunar injection.

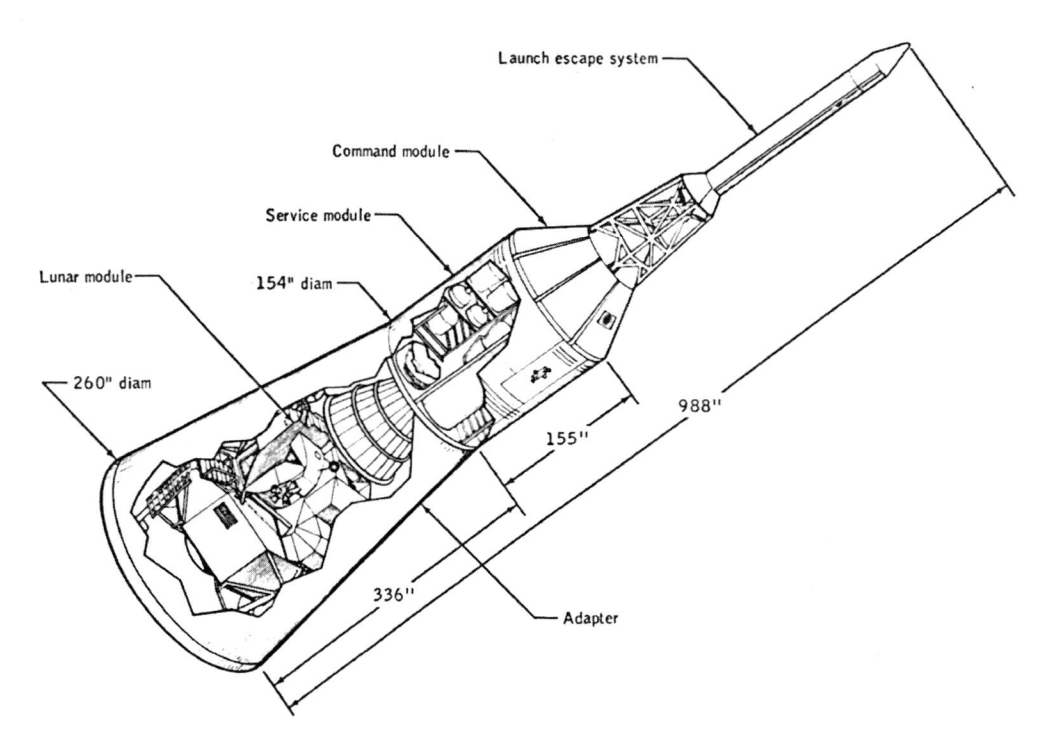

Figure XVI-1.- Apollo spacecraft.

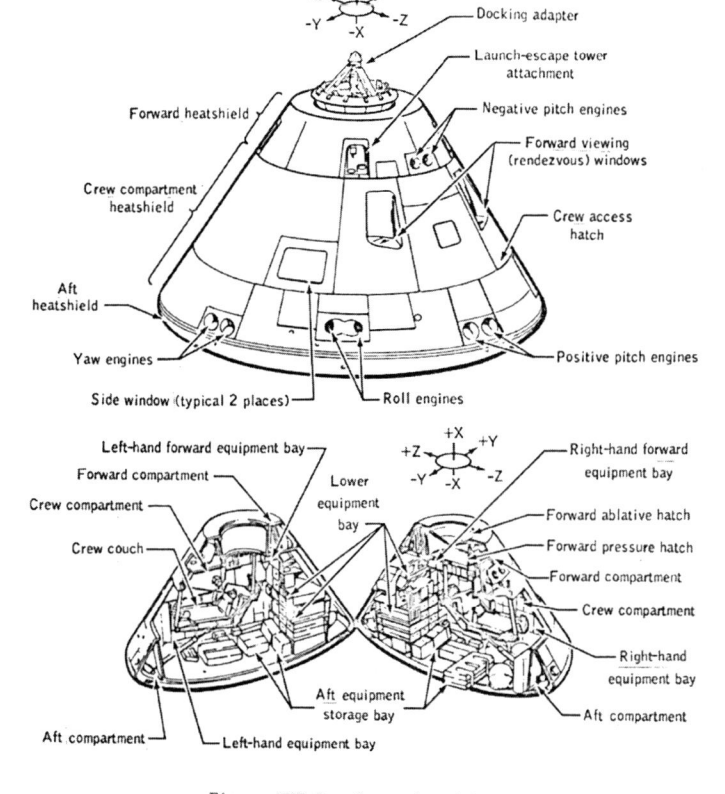

Figure XVI-2.- Command module.

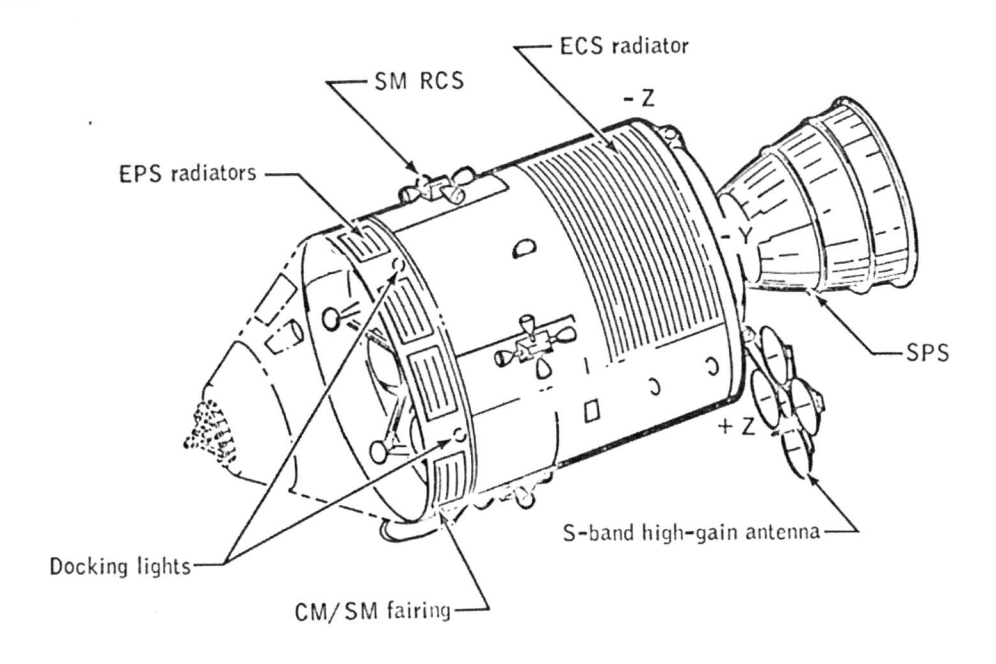

Figure XVI-3.- Service module.

SECTION XVII

UNMANNED LUNAR SATELLITES AND PROBES

Satellite Delivery by Apollo

The Apollo vehicle can carry a satellite to be deployed in lunar orbit. The empty bay 1 in the service module (SM) is adequate for housing a vehicle such as the 850-pound lunar orbiter or lighter satellites such as the anchored interplanetary monitoring platform (AIMP) or the Pioneer. The satellite capabilities are given in table XVII-I. At deployment, the skin over bay 1 is jettisoned, and the satellite is erected mechanically, spun-up for stabilization if necessary, and then ejected from the SM.

The advantages of orbital precision and success are obtained by utilizing the inherent flexibility of a manned vehicle to deliver a satellite to the desired space trajectory. In the delivery of either and AIMP satellite or a Pioneer satellite to a lunar orbit, high orbital precision is obtained from the primary navigation of the Manned Space Flight Network (MSFN), from the Apollo onboard secondary navigation, and from the in-flight Apollo targeting capability. The root-mean-square (rms) position and velocity uncertainties of the MSFN on Apollo are approximately 1 nautical mile in position and 4 ft/sec in velocity following one orbit of MSFN tracking.

Increased orbital sophistication and greater probability of success are also possible. The precision of the Apollo lunar orbit makes a compromise in desired orbital parameters unnecessary. A compromise is necessary for the AIMP satellite in order to increase the probability for success. Permutation of orbit; by a satellite from a precise lunar orbit permits much more radical ellipticity with less danger of losing the lunar orbit.

Landing Probes

Unmanned landing probes can be of various configurations (figs. XVII-1, XVII-2, and XVII-3). A probe can be sent to the lunar surface by direct flight from the earth as in the case of Surveyor, a probe can be released from lunar orbit as a hard lander, or a probe can be contained in a payload module that is delivered as part of an unmanned logistics vehicle (refs. 4 and 5).

Surveyor probe.- The Surveyor probe (fig. XVII-1) represents a class of vehicles designed to land and remain on the lunar surface. The probe is propelled to the moon by the Atlas-Centaur launch vehicle where the probe lowers itself to the lunar surface by means of automatically controlled rockets. The Surveyor weighs 2200 pounds and carries a payload of approximately 60 pounds of scientific instruments. Three Surveyors will be launched during the next year.

Hard lander probe.- The hard lander probe (figs. XVII-2 and XVII-3) can be released from lunar orbit to land at a predetermined lunar location. The hard landers weigh 300 to 400 pounds and can carry payloads of 70 to 250 pounds. The landers can withstand shock loads of more than 250g while landing at approximately 50 ft/sec. The operating life can be extended to approximately 6 months. Various instrumentation payloads are capable of operating from the capsule, and the data are transmitted back to earth or to an orbiting spacecraft. The capsule schedule, from the development to the delivery of the first flight unit, is approximately 3 years.

Soft lander probe.- A variety of soft lander probes (fig. XVII-4), such as a modified Surveyor or scaled-down device similar to the LM descent stage, can be developed. The devices could be released from lunar orbit as separate spacecraft to land at some predetermined lunar location. A variety of instrumentation systems could be carried on the devices, and the data would be transmitted back to earth.

Surface Crawlers

The objective of the lunar surface crawler (LSC) is to carry out an extended photovisual topographic survey with a detailed photovisual examination of interesting objects on the lunar surface (ref. 6).

A crawler, similar to one shown in figure XVII-5, is a remotely or automatically controlled surface vehicle, small enough to be carried to the moon on either a Surveyor or on the LM vehicle. After landing, the LSC is released from its carrier to undertake independent exploration of the lunar surface. The crawler weighs approximately 155 pounds and can carry a 25-pound payload. The payload could be a stereoscopic 500-line photofacsimile or television (TV) camera with transmission capabilities back to earth. The preferred power source is a radioisotope thermoelectric generator (RTG) which allows for cross-country exploration, limited only by line of sight, communications, and control with earth. The estimated surface speed of the LSC is one-half mile per hour, with capability to cross a 20-inch crevasse, climb up or down a 30-inch step, or up or down a hard surface ($Cf = 0.6$) slope of 30°.

TABLE XVII-I.- CAPABILITIES OF SATELLITES

	AIMP	Pioneer	Orbiter
Experiments	Off-the-shelf instruments 　Solar-cell damage 　Two magnetometers 　　Ness 　　Sonett 　Plasma probe 　Thermal ion 　High-energy particles 　Ionizing radiation 　Telemetry study 　Range and range rate	Magnetometer (ARC) Cosmic ray 　University of Minnesota 　GRCSW Plasma probe (ARC) Cosmic dust 　GSFC Electric field Radio propagation 　Stanford	Primary configuration 　TV 　Neutro albedo 　Cosmic ray 　X-ray 　Gamma ray 　　(6-ft boom) Secondary configuration 　IR detector 　Magnetometer 　Solar plasma 　TV
Availability	15 to 18 months after approval (includes gamma-ray experiment)	24 months after approval (includes propulsion system)	10 months after approval (Orbiter 6)
Lunar orbit desired	500 by 2000 km	128 km (80 n. mi.)	128 km (80 n. mi.)
Volume	15 ft^3	22 ft^3	125 ft^3
Satellite weight[a]	210 lb	145 lb (320 lb)[b]	850 lb
Orientation system	Sun sensors and gas jets	Sun sensors and gas jets	Complete system
Communications	Transmitter, 136.02 MHz Receiver, 148 MHz Bits/sec, 29.6	Transmitter, 2292 MHz Receiver, 2110 MHz Bits/sec, 512, 256, ..., 8	Transmitter Receiver Bits/sec

[a]The ejection mechanism estimated to be 20 percent of satellite weight and must be added to the satellite weight.

[b]The weight increases to 320 pounds with the propulsion system.

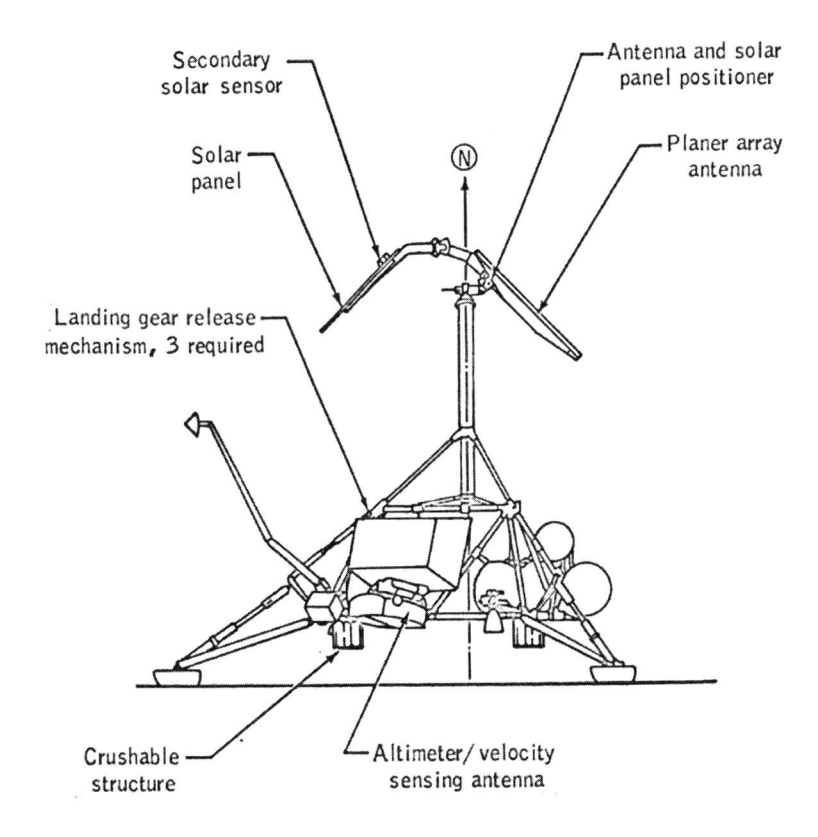

Figure XVII-1.- Surveyor probe.

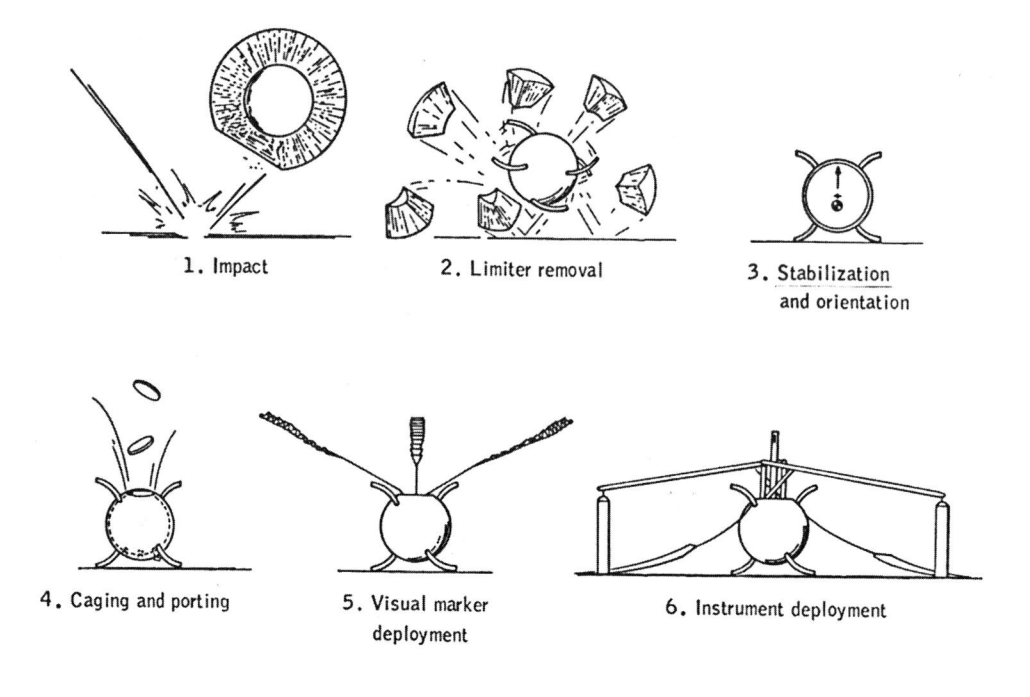

1. Impact
2. Limiter removal
3. Stabilization and orientation
4. Caging and porting
5. Visual marker deployment
6. Instrument deployment

Figure XVII-2.- Typical hard lander erection and leveling sequence, proposal I.

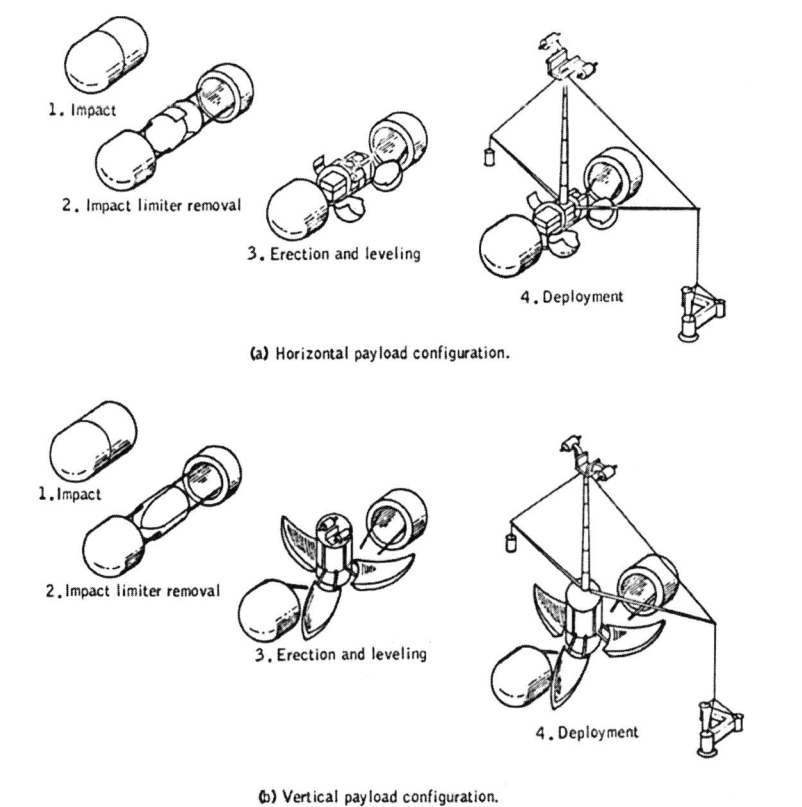

1. Impact
2. Impact limiter removal
3. Erection and leveling
4. Deployment

(a) Horizontal payload configuration.

1. Impact
2. Impact limiter removal
3. Erection and leveling
4. Deployment

(b) Vertical payload configuration.

Figure XVII-3.- Typical hard lander erection and leveling sequence, proposal II.

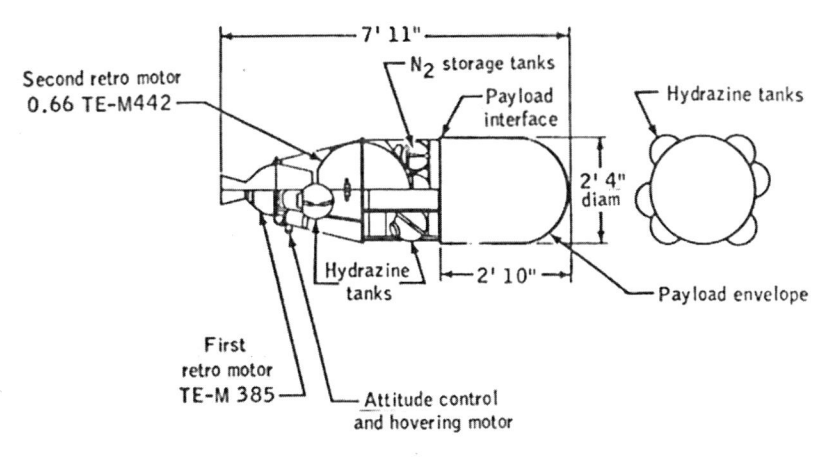

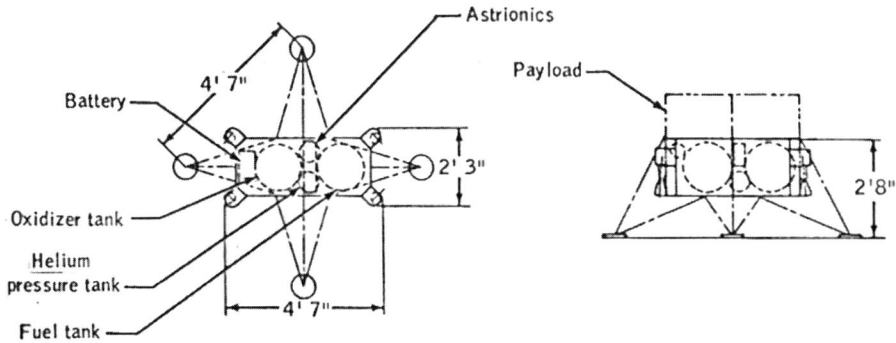

Figure XVII-4.- Soft lander probe.

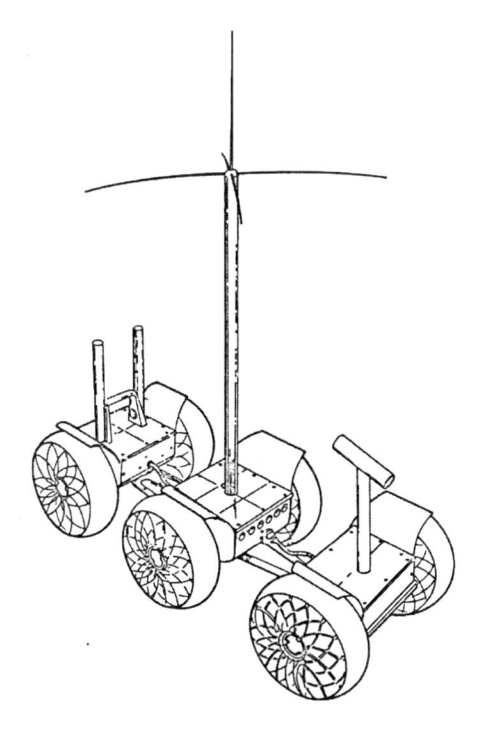

Figure XVII-5.- Unmanned lunar surface crawler.

SECTION XVIII

HARDWARE AND MISSION COSTS

Relative mission cost can be approximated by adding the appropriate nonrecurring and item costs from table XVIII-I for the hardware involved in a specific mission. It should be noted that certain nonrecurring costs are based on previous development of other hardware. If the previous item of hardware has not been developed, then its nonrecurring cost would have to be included in the mission being considered. For example, if a mission required. the use of a 7-day lunar module (LM) and the 3-day LM had not been developed, the nonrecurring cost for the 7-day LM would be $37.3 million, which represents $20.2 million for 3-day LM development (item 2), plus $17.1 million for development to extend the 3-day LM to 7 days (item 3). The number 2 in the fifth column indicates that the development cost for item 3 is dependent upon the previous development of item 2.

The figures for item cost shown in table XVIII-I include, in addition to item hardware cost, the cost of launch support and mission planning. Nonrecurring costs are the costs associated with development of an item of hardware and remain constant regardless of the number of items produced.

Tables XVIII-II, XVIII-III, XVIII-IV, and XVIII-V provide the relative costs of four groups of four potential lunar exploration missions where group 1 missions are all single launches with group 2, 3, and 4 missions utilizing a combination of single and dual launches. It is of interest to note that the nonrecurring landing vehicle costs are relatively low when compared to recurring costs, particularly for dual-launch missions.

TABLE XVIII-I.- NONRECURRING AND ITEM COST SUMMARY

Item	Component	Life, days	Nonrecurring cost, millions of dollars	Nonrecurring cost is delta above item number	Item cost, millions of dollars
1	Lunar module (LM)	--	--	--	36.44
2	Extended lunar module (ELM)	3	20.22	1	36.99
3	ELM	7	17.15	2	40.54
4	ELM storable	90 + 7	41.73	2,3	43.10
5	LM taxi	3	34.74	2	39.89
6	LM shelter	90 + 14	119.13	5	41.85
7	Lunar payload module (LPM)	--	19.82	1	27.51
8	Augmented lunar module (ALM)	14	113.55 93.33	1 2	40.54
9	ALM shelter	90 + 14	105.80	8	41.85
10	Modular LM truck		37.3	1	26.00
11	Modular ALM truck		37.3	8	26.00
12	LM truck shelter (LMTS)	90 + 14	107.4 70.1	1 10	39.55 39.55
13	Mission extension provisions	--	17.0	--	3.50
14	Command and service module (CSM)	--	--	--	47.5
15	Saturn V launch vehicle	--	--	--	148.0
16	Local scientific survey module (LSSM)	--	40.0	--	4.0
17	Lunar flying unit	--	13.0	--	1.5
18	Manned flying system (MFS)	--	48.0	--	3.0
19	Mobile exploration (MOBEX) unit	--	350.0	--	18.0
20	Surveyor	--	--	--	11.1
21	Hard lander	--	50.0	--	3.9
22	Soft lander	--	83.3	--	12.0
23	Drill	--	15.1	--	.9

TABLE XVIII-II.- GROUP 1 MISSIONS ($1095.7 MILLION)

Mission components	Cost, millions of dollars	
	Nonrecurring	Recurring
Mission 1A		
ELM (3 day)	20.2	37.0
Flying vehicle, 2	13.0	3.0
Saturn V	--	148.0
CSM	--	47.5
Subtotal	33.2	235.5
Mission 1B		
ELM (3 day)	--	37.0
Flying vehicle, 2	--	3.0
Saturn V	--	148.0
CSM	--	47.5
Subtotal	--	235.5
Mission 1C		
ALM (14 day)	113.5	40.5
Flying vehicle, 2	--	3.0
Saturn V	--	148.0
CSM	--	47.5
Subtotal	113.5	239.0
Mission 1D		
ALM (14 day)	--	40.5
Flying vehicle, 2	--	3.0
Saturn V	--	148.0
CSM	--	47.5
Subtotal	--	239.0
Total	146.7	949.0

TABLE XVIII-III.- GROUP 2 MISSIONS ($1513.1 MILLION)

Mission components	Cost, millions of dollars	
	Nonrecurring	Recurring
Mission 2A		
ELM (3 day)	20.2	37.0
Flying vehicle, 2	13.0	3.0
Saturn V	--	148.0
CSM	--	47.5
Subtotal	33.2	235.5
Mission 2B		
ELM (3 day)	--	37.0
Flying vehicle, 2	--	3.0
Saturn V	--	148.0
CSM	--	47.5
Subtotal	--	235.5
Mission 2C (dual launch)		
ELM (7 day)	17.1	40.5
LPM	19.8	27.5
Flying vehicle	--	3.0
LSSM, 1	40.0	4.0
Saturn V, 2	--	296.0
CSM, 2	--	95.0
Subtotal	76.9	466.0
Mission 2D (dual launch)		
ELM (7 day)	--	40.5
Flying vehicle, 2	--	3.0
LPM	--	27.5
LSSM, 1	--	4.0
Saturn V, 2	--	296.0
CSM, 2	--	95.0
Subtotal	--	466.0
Total	110.1	1403.0

TABLE XVIII-IV.- GROUP 3 MISSIONS ($1627.5 MILLION)

Mission components	Cost, millions of dollars	
	Nonrecurring	Recurring
Mission 3A		
ELM (3 day)	20.2	37.0
Flying vehicle, 2	13.0	3.0
Saturn V	--	148.0
CSM	--	47.5
Subtotal	33.2	235.5
Mission 3B		
ALM (14 day)	113.5	40.5
Flying vehicle, 2	--	3.0
Saturn V	--	148.0
CSM	--	47.5
Subtotal	113.5	239.0
Mission 3C (dual launch)		
ALM (14 day)	--	40.5
Flying vehicle, 2	--	3.0
ALM truck	37.3	26.0
LSSM	40.0	4.0
Saturn V, 2	--	296.0
CSM, 2	--	95.0
Subtotal	77.3	464.5
Mission 3D (dual launch)		
ALM (14 day)	--	40.5
Flying vehicle, 2	--	3.0
ALM truck	--	26.0
LSSM	--	4.0
Saturn V, 2	--	296.0
CSM, 2	--	95.0
Subtotal	--	464.5
Total	224.0	1403.5

TABLE XVIII-V.- GROUP 4 MISSIONS ($1675.2 MILLION)

Mission components	Cost, millions of dollars	
	Nonrecurring	Recurring
Mission 4A		
ELM (3 day)	20.2	37.0
Flying vehicle, 2	13.0	3.0
Saturn V	--	148.0
CSM	--	47.5
Subtotal	33.2	235.5
Mission 4B		
ELM (3 day)	--	37.0
Flying vehicle, 2	--	3.0
Saturn V	--	148.0
CSM	--	47.5
Subtotal	--	235.5
Mission 4C (dual launch)		
LM shelter	119.1	41.8
LSSM	40.0	4.0
ELM taxi	34.7	39.8
Flying vehicle, 2	18.0	3.0
Saturn V, 2	--	296.0
CSM, 2	--	95.0
Subtotal	211.8	479.6
Mission 4D (dual launch)		
LM shelter	--	41.8
LSSM	--	4.0
ELM taxi	--	39.8
Flying vehicle, 2	--	3.0
Saturn V, 2	--	296.0
CSM, 2	--	95.0
Subtotal	--	479.6
Total	245.0	1430.2

REFERENCES

1. Saturn Apollo Applications Program Definition at KSC. (Contract No. NAS 10-3678), January 1967.

2. Environmental Test Methods for Aerospace and Ground Equipment. MIL-STD-810A (USAF), June 23, 1964.

3. Electromagnetic Interference - Test Requirements and Test Methods. MIL-STD-826A (USAF), June 30, 1966.

4. Study of Science Capsules for the Lunar Survey Probe. (Contract No. NASw-1138), Aeronutronic, Division of Philco Corp., a subsidiary of Ford Motor Company, February 25, 1966.

5. Lunar Survey Probe Survivable Capsule Study, Final Report-Phase B. (Contract No. NASw-1134), Space General Corp., May 14, 1965.

6. Surveyor Block II Mission Study. JPL Engineering Planning Document No. 259, April 1965.

BIBLIOGRAPHY

Apollo Operations Handbook. Block II Spacecraft, Vol. I Spacecraft Description. SM2A-03-Block II-(1) (Contract No. NAS 9-150), March 1, 1967.

Apollo Operations Handbook. Vol. 1, Lunar Module. LMA790-3-LM 2 (Contract No. NAS 9-1100), Grumman Aircraft Engineering Corp., January 1, 1967.

Bi-Monthly Progress Report for March-April, 1967. 378C-16 (Contract No. NAS 9-6608), Grumman Aircraft Engineering Corp., May 15, 1967.

Candidate LM Derivatives. Sections 2.1, 2.2, and 2.3. 378C-17 (Contract No. NAS 9-6608), Grumman Aircraft Engineering Corp., June 1967.

Early Lunar Shelter Design and Comparison Study, Final Report. (Contract No. NAS 8-20261), Garrett AiResearch. Corp., February 8, 1967.

LSSM for Apollo Applications Program, Final Report. BSR 1495 (Contract No. NAS 8-20378), February 1967.

Lunar Staytime Extension Module, Final. Report. GER 12246 (Contract No. 1-4277), Goodyear Aerospace Corp., August 21, 1965.

Lunar Surface Mobility Systems Comparison and Evolution (MOBEV), Final Report. BSR 1428 (Contract No. NAS 8-20334), November 1966.

Specified LSSM Design Study, Technical. Report. D2-113471-2 (Contract No. NAS 8-20340), November 1966.

Study of Mission Modes and System Analysis for Lunar Exploration (MIMOSA). Vol. II, MIMOSA Planning Methodology, Parts 1 and 2. LMSC-A847943 (Contract No. NAS 8-20262), April 30, 1967.